WE FED AN ISLAND

An ANTHONY BOURDAIN BOOK
ecco

An Imprint of HarperCollins*Publishers*

WE FED AN ISLAND

THE TRUE STORY OF REBUILDING PUERTO RICO, ONE MEAL AT A TIME

JOSÉ ANDRÉS
with RICHARD WOLFFE

100% of the author's net proceeds will be donated to the Chef Relief Network of World Central Kitchen for efforts in Puerto Rico and beyond.

A hardcover edition of this book was published in 2018 by Ecco, an imprint of HarperCollins Publishers.

FIRST ECCO PAPERBACK EDITION PUBLISHED 2019.

Foreword © 2018 by Lin-Manuel Miranda and Luis A. Miranda, Jr.
Photograph on pages ii–iii by Felix Lipov/shutterstock
Photograph on page vi by MaxyM/shutterstock
All photo insert images courtesy of World Central Kitchen

Designed by Renata De Oliveira

Library of Congress Cataloging-in-Publication Data has been applied for.

ISBN 978-0-06-286449-9

19 20 21 22 23 RS/LSC 10 9 8 7 6 5 4 3

To the unknown heroes: the chefs,
the volunteers, the military, the first
responders, and all the forgotten people
around the world who put their lives
aside to feed others in need.

CONTENTS

FOREWORD

THREE DAYS. THAT'S HOW LONG IT TOOK FROM THE TIME HURRICANE
Maria hit Puerto Rico to when we knew our family was still alive
thanks to a Facebook picture a cousin had posted. Five days to
hear if our family was fine (they were) and to hear if the house
our family had built years ago had survived (it did not).

It's a horrible feeling—waiting and not knowing. You feel help-
less. You worry. You imagine the worst. You feel like there is nothing in your skill set that can help your loved ones who are so far
away. We got relief from that worry in only five days. We were
fortunate. It was a much longer wait for many others.

Now imagine a different kind of waiting and not knowing.
Waiting without power. Waiting for food. Waiting for running water. Not knowing if your family one town over is alive. Not knowing
if the roads are clear or if it's safe to go in search of supplies. Not
knowing if the hospitals will continue to function. Not knowing
if anyone from the mainland is on their way with relief or if they
even have a strategy in place to help you. Not knowing if your
entire island has been forgotten.

What you are about to read is the story of someone who
helped people in a time of crisis. Someone who did not wait for

paperwork or permits while people starved. Someone who saw that without nourishment, no one would have the strength to re-build and recover in the long days, months, and years ahead.

Chef José Andrés was one of the first people to arrive on the island for humanitarian purposes after Maria hit. He faced obsta-cle after obstacle as he tried—and succeeded—in setting up what became the biggest kitchen with "real" food feeding Puerto Ricans for two months after the storm. His first day of operations, Chef José Andrés and Chefs for Puerto Rico, a cadre of volunteer chefs from the island, prepared 1,000 meals to feed Puerto Ricans. By the second day, they doubled their output. By three months, 3 mil-lion meals were prepared and served throughout the island.

José Andrés's mission in Puerto Rico seemed simple: develop a meal program for Puerto Ricans-in-need, with delicious food that was cooked and delivered on the same day, using local prod-ucts to stimulate the local economy. His clarity of purpose put him at odds with almost every large institution involved in relief efforts on the island. He did not want to hear about bidding pro-cesses, meetings, or excuses about why scaling up could not be done. And he did not take NO for an answer. We were amazed at what he was accomplishing in the aftermath of the storm and could not wait to get down there and lend a hand.

Upon arriving in Puerto Rico, the first thing we noticed was a once-green country that now looked as if it had been set on fire: all vegetation was gone or dead. We could see the devastation, the FEMA blue tarps as makeshift roofs everywhere, the darkness descending as the sun set. The sea of desperation and need was best summarized by the mayor of San Juan's outcry, "We are dy-ing here" in a plea to the federal government for help. Amongst all this, Chef José Andrés had created a beacon of sustenance, an oasis in every place from where they were cooking, including the Choliseo.

When we arrived in Puerto Rico, the first stop was to join the assembly line at Chef José Andrés's kitchen. As we walked in we were overcome by the joy of the volunteers, by the aroma of Puerto Rican cuisine, by the enormity of the effort. As sandwich-making beginners our instructions were simple—lots of mayo, cheese, and ham. Though the sandwiches seemed simple to many, for Chef José Andrés, they were magical: "I have created many avant-garde dishes as a chef but there are few meals I'm prouder of than the hundreds of thousands of sandwiches we made in Puerto Rico."

For more experienced workers, the cooking menu had expanded beyond sandwiches to include *sancocho,* a stew including a variety of meats, corn, and vegetables. As a way to deliver large quantities of calories, nutrition, and comfort to many storm survivors at a time, it's hard to beat *sancocho.* As José says, "When you eat *sancocho,* you think of your grandmother and it puts a smile on your face." And of course, *arroz y habichuelas, arroz con pollo,* paella, added to a menu that, though created at a time of crisis, tasted and looked as if it came from your mother's kitchen.

What Chef José Andrés and his team were able to achieve in a short time seems unbelievable—producing tens of thousands of fresh meals each day for Puerto Ricans in need—but what is so important to remember is that it was *possible.* Despite the challenges, Chef Andrés made it happen. There is so much more we can be doing collectively and we have to expect better from our government. It is a national embarrassment that a year later there are areas of Puerto Rico that have never had power restored and vast areas of the island experience intermittent blackouts. The people of Puerto Rico cannot be forgotten in this ongoing time of need. We are thankful to Chef Andrés for all he has done and all he continues to do. We are proud of the many who volunteered, donated, and raised their voices. We will forever be in debt to those, like Chef José Andrés, who put their lives on hold to help Puerto Rico.

This book is the Chef's story, a remarkable one. We'll let him tell it.

Siempre,
Luis A. Miranda, Jr. and Lin-Manuel Miranda

WE FED AN ISLAND

PROLOGUE

THE FIRST TIME I TRAVELED TO THE CARIBBEAN WAS BY SHIP. JUST LIKE the early colonial explorers, I sailed into Santo Domingo and marveled at its beauty and geography. I was a young man, serving briefly in the Spanish navy on the *Juan Sebastián de Elcano,* a majestic, four-masted topsail. The third-largest tall ship in the world, it was named after the Spanish explorer who was the captain of Magellan's fleet, and the first man to circumnavigate the world. Although I had no idea at the time, this was the start of my love affair with the Caribbean and with America.

So a few years later, when I was drifting between jobs for several months, I could only say yes to the chance to work in Puerto Rico and to return to this magical sea. I was a young chef, learning my trade shortly before settling down in Washington, D.C., to start my restaurants and my family. But I will never forget the sights and sounds of those weeks cooking at La Casona in the Santurce neighborhood of San Juan: the spirit of the salsa, the nightly call of the *coquí* frogs, the lush green of the tropical leaves.

Many decades later, as an established chef with many restaurants to my name, I returned to revive one of the jewels of the island's glamorous heyday in the 1950s and 1960s: the magical place known as Dorado Beach. My restaurant, Mi Casa, is part of the former home of the visionary American who preserved the natural beauty of this northern shore, Clara Livingston. She sold her plantation to Laurance Rockefeller, the environmentalist, who carefully developed it as one of his RockResorts, turning it into a refuge for Hollywood stars and American presidents. I was honored to be part of its revival, and my work there created lifelong friendships with many of the island's chefs, its entrepreneurs and the everyday Puerto Ricans who embody its creative and welcoming spirit.

So when Hurricane Maria devastated Puerto Rico in late September 2017, it felt like destiny was driving me back to the place where it all began for me. It was as if two timelines were meeting at the same point in the warm Caribbean Sea: my past and my present, this island's Spanish roots and its American identity. The echoes of its history merged with the urgent voices of today's crisis. I felt I belonged here because my ancestors were not so different from the settlers who fought and farmed and cooked here for so many centuries before I arrived. Puerto Rico is the perfect mix of Spanish and American. It's the perfect mix of my culture. There are African Americans here. They have the blood of my people and the blood of the Africans who were forced to come here.

How could I not be here?

AS PEOPLE SANG ALONG TO "DESPACITO" THROUGH THE SUMMER OF 2017, how many of them understood that Luis Fonsi's hit was born in this American-Spanish corner of the Caribbean? If you were going to create a song that represented the perfect blend of cultures to break through the language barrier, a song that would garner the most views ever on YouTube, it would be right here in Puerto Rico. And when the hurricanes landed just a few weeks after the end of summer vacation, how many of those "Despacito" fans had any idea the islanders were American citizens just like them?

These islands are not just tourist destinations or hurricane targets. They are the first places the original colonists exploited and reshaped in their own image. They bear the scars of their abuse and neglect to this day. We cannot value Puerto Rico simply for its crops or the national security advantage it offers, and then ignore its inhabitants when they need our investment to break the cycle of poverty or to recover from nature's fury.

To understand our responsibilities, we need first to understand our history here. That includes the unique contribution this part of the world has played in our American success. After all, it was a hurricane in 1772 that brought Alexander Hamilton from nearby St. Croix to New York, where he would change the course of this nation and the world. Then just seventeen years old, working as a clerk for a business on the island that traded with America, Hamilton penned a letter to his father that was so well written that a group of wealthy islanders raised the money to send him away for his education.[1] His letter was a plea for compassion and disaster relief. "O ye who revel in affluence," he wrote, "see the afflictions of humanity and bestow your superfluity to ease them. Say not, we have suffered also, and thence withhold your compassion. What are your sufferings compared to those? Ye have still more than enough left. Act wisely. Succour the miserable and lay up a treasure in Heaven."[2]

It was clear to everyone on Puerto Rico that the president himself knew nothing of America's history on this island before the hurricane struck. When Donald Trump mocked the pronunciation of the island's name, he recalled a time when Americans ruled without regard to its identity. "We love Puerto Rico," he told a crowd of supporters at the White House for National Hispanic Heritage Month, barely two weeks after the hurricane. "Puerto Rico," he repeated, emphasizing the Spanish accent once again. "And we also love Porto Rico," he added, laughing at his own joke.[3]

Our response to a natural disaster has never depended on a

person's accent or politics. We may be Republicans or Democrats—or apolitical, for that matter—but we are fundamentally all Americans. This country has a long and proud tradition of taking care of Americans, and non-Americans, in their moment of need.

THERE'S SOMETHING FUNDAMENTAL ABOUT FOOD; ABOUT PREPARING, cooking and eating together. It's what binds us; it's how we build community. Eating isn't functional. Food relief shouldn't be either. Whether I am cooking for Washingtonians or refugees, my job as a chef is the same: to feed the many. Whether I am creating an avant-garde meal that deconstructs your idea of a familiar meal, or a giant pot of rice and chicken that fills your belly, I believe in the transformational power of cooking.

A plate of food is much more than food. It sends a message that someone far away cares about you; that you are not on your own. It's a beacon of hope that maybe somewhere, something good is happening. It's the hope that America will become America again. That is what a plate of food is. It's a message from every man and woman on my team saying that we care, that we haven't forgotten, and it allows those in despair to have a little bit more patience, for one more day.

As I developed my vision for a new model of food relief, I learned a profound lesson from my mentor Robert Egger, who is America's leading advocate on food issues. "Too often," he said, "charity is about the redemption of the giver, not the liberation of the receiver." I do believe that food relief should help liberate the receiver, and that far too often, it has been defined and delivered to redeem the giver. We need to build a new model of disaster relief and food aid that understands the needs and desires of the receiver, and we need to do that right now.

We achieved something extraordinary in Puerto Rico, preparing more than 3 million meals as a small nonprofit, while the federal government and the giant charities struggled to get anything done. We overcame blocked roads and collapsed bridges, political opposition and bureaucratic red tape, supply bottlenecks and cash crunches. It was hot, sweaty, exhausting work. But it was also life-changing and inspiring, channeling our love to do something as simple as this: to feed the people.

Although each disaster is different and each one is complex, the priorities are simple. There is no recovery to manage, and no citizens to govern, if we cannot get water and food to the people. And yet, if you ask around—and believe me, I did—there is nobody, and no single organization, in charge of feeding the people. The experts tell me that everyone is in charge, but what I have seen is that means nobody is in charge. Food relief is not just a question of results and accountability. It is a moral necessity. As Tom Joad says in Steinbeck's classic from the Great Depression, *The Grapes of Wrath,* "Wherever they's a fight so hungry people can eat, I'll be there."

This is a story of our fight so hungry people could eat. We didn't feed them as much as we wanted. But we were there, even though we were never supposed to be.

CHAPTER 1

LANDFALL

MARIA EXPLODED TWO DAYS BEFORE SHE ARRIVED IN PUERTO RICO.
Over the course of just twenty-four hours, her winds doubled in speed from 80 to 160 miles per hour. The next day, she ripped through the island of Dominica as a Category Five hurricane: the first on record to do so. She weakened a little as she ripped the roofs off almost every building, tore out almost every electrical and telephone pole, stripped the leaves off almost every tree, crushed the banana crops and killed the livestock. No one was spared, not even the island's prime minister, Roosevelt Skerrit. "My roof is gone. I am at the complete mercy of the hurricane. House is flooding," he posted on Facebook, just before he was rescued from his official residence.[1]

Shortly before sunrise the next day, Maria landed as a Category Four hurricane on the southeast coast of Puerto Rico. Her center was 50 to 60 miles across, or about half the length of the main island, and her winds blew as fast as 155 miles per hour. She slashed and tore westward on a diagonal path across the beaches and mountains, the villages and the cities, the farms and the luxury apartments. Maria took her time in devastating anything exposed to the elements, lumbering along at just 10 miles an hour. She snapped apart huge wind turbines, plucked up the electric grid and tossed aside solar panels. She silenced the cell phone towers, uprooted the old telephone poles, and flicked over weather radar and satellite dishes. She clawed out the forests on the hillsides, and left only the naked trunks of the trees she spared. She heaved the sea into low-lying homes, and forced high, raging floods through mountain ravines. She destroyed the coffee farms, decimated the dairy herds and demolished the greenhouses. She darkened the hospitals and soaked the wards with rainwater. What her sister Irma had weakened with a glancing blow, less than two weeks earlier, Maria finished off with a direct hit.

For the next two days, stunned Puerto Ricans struggled to survive the onslaught of catastrophic rain and flooding. They rescued their neighbors and gathered together their food and clean water. They began to dig their way out: heaping household debris into piles on the streets, cutting paths through fallen trees to open roads and driveways, carefully treading around or moving the wires and cables that now lay on the ground. As they began to clear out, the morgues began to fill up. At first the bodies were those of the direct victims of the winds and floods. But soon, with most of the hospitals dark and damp, they were of the elderly and the sick who died at home, or in senior homes or at the stricken medical centers. News organizations estimated the number of dead at more than a thousand, but nobody knew for sure. At the Institute of Forensic Sciences in San Juan, they would need eleven refrigerated trailers to hold all the bodies.[2]

The day after Maria, Donald Trump was under no illusion how catastrophic the damage was. "Puerto Rico was absolutely obliterated," he told reporters after a meeting at the United Nations. "Puerto Rico got hit with winds. They say they have never seen winds like this anywhere. It got hit as a Five—Category Five storm—which just literally never happens. So Puerto Rico is in very, very, very tough shape. Their electrical grid is destroyed. It wasn't in good shape to start off with, but their electrical grid is totally destroyed. And so many other things. So we're starting the process now and we'll work with the governor and the people of Puerto Rico.

"So Puerto Rico will start the process . . . We're going to start it with great gusto. But it's in very, very, very perilous shape," he concluded. "Very sad what happened to Puerto Rico."[3]

That night, Trump flew to New Jersey to spend the long weekend at his golf club. He and his aides didn't mention Puerto Rico in public again, but they found the time for a campaign trip to Alabama. While at the golf club, Trump held a meeting with several of his cabinet officials, including his Homeland Security secretary. But the topic was his Muslim travel ban, not the hurricane. Trump's staff would not say if he spoke to anyone about Puerto Rico through the four-day weekend. But it was clear from his Twitter activity that he was focused on at least four issues: attacking NFL players for their protests during the national anthem, attacking Senator John McCain for his vote against repealing Obamacare, attacking the North Korean leader Kim Jong-Un and attacking the news media.[4]

THE NEWS FROM PUERTO RICO WAS FRUSTRATINGLY SKETCHY. I KNEW there was a crisis but it was hard to assess without being on the ground. Most of the island's cell phone towers—around 85 percent of the 1,600 towers on Puerto Rico—were down.[5] Nobody could find a working Internet or phone connection. After two days of trying to understand the situation, I knew I had to catch the first flight out there. By Saturday, three days after Maria ravaged the island, San Juan airport was only open to military flights. I booked seats on flights, but nothing was moving. I tried to get my

hands on a satellite phone, tweeting to the world to see if someone could lend me one. But it wasn't easy on a weekend, even in
Washington, D.C. I called my friend Nate Mook, whose documen-
tary work had taken him across the world and who knew much
more about satellite phones than I did. Nate had produced my
PBS show, Undiscovered Haiti, and he knew what I meant by the
power of food to rebuild lives. Like so many times before, I didn't
have a clear plan in mind, but I wanted to see what was happening.

"I'm going to bring some cash and solar lamps," I told him.
"What are you doing? Do you want to come?"

"Yeah!" he shot back.

We knew that downed communications and electricity would
make life difficult, but Puerto Rico was still the United States. It
couldn't be as bad as Haiti. We thought we'd be back by the end
of the week.

We were wrong.

The next day, Sunday, marked the first day the White House
had any contact with a Puerto Rican leader. Vice President Mike
Pence called Jenniffer González-Colón, the island's non-voting
member of the House of Representatives. For three days, Donald
Trump had said nothing in public, not even a tweet, about the hurricane or its impact on the island. In fact it was Hillary Clinton
who was the first leader to make a public statement, on that Sunday, as she tweeted to Trump and Defense Secretary James Mattis
to send the hospital ship USNS Comfort to Puerto Rico. "These are
American citizens," she implored them, posting a link to photos
of islanders wading through waist-deep waters to move through
their own streets. Her tweet was liked more than 300,000 times.

It was the first day a commercial flight landed in San Juan: a
single Delta flight. Every other flight ended in failure and simply
turned back.

I was following the news nervously, and I knew I needed to
be there. Watching CNN, I only had to look at my wife, Patricia,

for her to know what I was thinking. We drove to the REI store to buy solar lamps, water purification pills and survival gear for the hurricane victims, but we really didn't know what to expect. I just wanted to avoid becoming a problem in a place where people were suffering already. One of our biggest priorities was gathering cash for the trip to buy supplies. Between my wife's ATM card and my own, I managed to get my hands on $2,000. My executive assistant Daniel Serrano brought me another $1,500.

I managed to make brief contact with my friend José Enrique Montes, whose small restaurant in Santurce was home to some of the very best food in Puerto Rico. His business was wrecked, with no power and a leaking roof. With his refrigerated food going to waste, in a neighborhood full of hungry people, he did what chefs do: he started cooking. True to his roots and talents, he made the hearty, tasty soup known as *sancocho*.

Somewhere between a stew and a thick soup, *sancocho* is the Caribbean version of the Spanish *cocido*, brought to the region via the original colonial settlers who passed through the Canary Islands. In the Canaries, the last stop in European territory before the trade winds carried the ships to the Caribbean, *sancocho* was made with fish. By the time the dish became a favorite of the Caribbean and Latin America, it had shifted to a meat-based stew, often featuring lots of different meats, made with corn and a mix of vegetables. As a way to deliver calories and comfort to storm survivors in large quantities, it was hard to beat *sancocho*. "When you eat *sancocho*, you think of your grandmother and it puts a smile on your face," says José Enrique.

We booked two flight options for Monday, just in case one of them collapsed. Nate and I had seats on an 8:00 a.m. Delta flight from New York's JFK Airport direct to San Juan, as well as a Spirit Airlines flight from Baltimore that passed through Fort Lauderdale. We thought about taking an Uber car from D.C. to New York, but chose instead to go to Baltimore for the flight through Florida. We figured that if the flight was canceled, we could always travel to Miami, where I have two restaurants.

At the airport in Fort Lauderdale, we made a beeline for the ATM machine. The news suggested the Puerto Rican banks were a long way from re-opening, so I needed more cash. But I couldn't remember what my PINs were, and my cards weren't working. I called up Patricia back home for help. Fortunately she is the organized and sensible person in my family. With her guidance, I got my hands on another $2,000, which I withdrew in four transactions of $500. The ATMs were not exactly set up for our heavy needs.

Inside the terminal we watched the news on the airport screens. It was not promising for our journey: San Juan's airport had lost power. Travelers were stranded inside, in the sweltering heat, sleeping on the floor while waiting for the power and the flights to return. The situation seemed desperate: no food, no water, no air-conditioning, no flights. People were prepared to suffer all that in the hope of getting a seat on the first flight off the island. How bad were the conditions at home for them to do that?

I tried to call José Enrique but the calls weren't connecting. I contacted instead one of my Puerto Rico partners to see if he could help set things up for my arrival. Kenny Blatt was one of the investors who helped revive the great Dorado Beach resort, transforming it into the oasis it is today, after decades of decline. My restaurant there, Mi Casa, was one of the jewels of my Think-FoodGroup businesses. Kenny was in touch with Alberto de la Cruz, the smart entrepreneur who runs Coca-Cola's bottling operations in Puerto Rico. Alberto let us know that the Puerto Rico governor had put Ramón Leal, the head of the island's restaurant association, ASORE, in charge of all kitchens on the island. Leal had been working with the governor on a feeding plan for the island since Hurricane Irma, two weeks earlier.

Our plane was full of worried families trying to rush back to check on their loved ones, or their property, or both. With the communication systems stricken, there was no practical way to find out if family members were alive and well, or to find out if homes were still under water. Despite all the uncertainties of air travel onto an island with no power, the many risks were outweighed by the even greater worries. The only way to be sure was to show up in person.

For me, this was the start of the challenge of a lifetime. Our plane was one of the first commercial flights to make it into San Juan after the hurricane. We had no idea what to expect and it seemed like the pilot didn't either. As we sat on the tarmac at Fort Lauderdale, he came out from his cockpit to ask if anyone had a satellite phone they could lend him. The passenger behind us said he did, but it was in his checked luggage. I sorely wished I had found that satellite phone back in Washington.

"We might have to get your checked bag out," the pilot said. "Once we are on the ground, we might need to talk to the air traffic control tower with the satellite phone so we can taxi over." There was no way to know if the controllers at San Juan airport would have power when we landed. We waited another forty-five minutes while the pilot located another satellite phone from a different Spirit flight. I couldn't believe the airlines were so unprepared for this kind of emergency.

Between the stress of the unexpected and the late-night packing and preparations, we were exhausted before the flight took off. But that didn't stop us mapping out our plans. We talked about my nonprofit World Central Kitchen: about the current state of the food operations in Haiti that Nate had filmed, as well as my recent experience in Houston, post-Harvey, where I saw firsthand how food relief on the mainland was hampered by old ways of thinking and inefficiencies. We envisioned an island-wide operation in Puerto Rico that was far more ambitious. We needed a robust technology platform that could handle multiple food requests and manage our supplies. We needed to be able to track those requests and the deliveries, as well as manage the donations we hoped would arrive. I dreamed of a system where people could text a website with the food request: maybe a shelter needed four hundred meals, and the system would locate the nearest kitchen that could help cook those meals. It was going to be a localized approach, with World Central Kitchen as the clearinghouse with the best technology. We were dreaming big dreams because the desperation seemed so overwhelming. You should never feel guilty about feeling ambitious when you're trying to help other people. If you don't dream, then reality will never change.

As our plane approached San Juan, there was devastation as far as the eye could see. Roofs were ripped off, with so many homes peeled open like tin cans. Trees were toppled for miles on end, or stripped of every single leaf. The trunks and limbs were so bare, Puerto Rico looked less like a tropical island and more like winter in my beloved home state of Maryland.

I texted Ramón as soon as I landed. The phone signal didn't seem to work, but some data was finding its way through. "We welcome you with open arms!!" he shot back, telling me to come directly to San Juan's convention center, where the government was headquartered, before we toured a couple of kitchens.

The airport was eerily quiet. There were no planes coming and going, no supply trucks busying themselves on the tarmac. Inside the terminal, there were no lights and no sounds. People seemed to be suffering in silence, without food or water. I immediately reached for my phone to tweet at my contacts, telling them to send food trucks to the airport.

We had booked a car from Europcar but discovered their location was off-site. So we walked up to the Avis counter and hoped for the best. I was lucky. One of the Avis staff recognized me from my cooking show on Spanish TV. That helped me talk him into renting us a precious Jeep that could travel the messy roads.

"If you need anything, come back and I'll help out," said my Avis friend.

"If I run out of gas, I don't think I can do that," I replied, only half joking.

As we drove out of the airport, it was clear that we needed the Jeep. The major roads were still strewn with dangerous debris: electric and telephone poles were lying where they had fallen, with their cables snaking alongside tree trunks and branches. Driving was a test of skill and nerves on an unpredictable obstacle course, in lanes that were suddenly blocked, and at intersections where there were no lights to control the traffic.

We headed straight for the convention center and parked on the side of the building alongside the Homeland Security Jeeps. There was a side door propped open with TV cables leading to the satellite trucks outside. We walked right in, and headed for the second floor, where government officials were supposed to be working on disaster relief in the many meeting rooms. Nobody stopped to ask us what we were doing there.

My friend Ramón Leal had told me about the biggest meeting, which was dealing with the most pressing issue: gasoline. We walked into the session and made ourselves at home. The room overlooked one of the halls that had been converted into a giant staging post for supplies, along with cots for officials to sleep in.

In our meeting, a group of business leaders were doing what the private sector does so well: solving the market's problems. Puerto Ricans were lining up for several hours every day to get a precious few gallons of gas for their cars, and clogging up the roads. The gas lines were the most visible sign of an economy that had ground to a halt. For the sake of individuals and businesses, these leaders needed to restore the fuel supply chain as rapidly as possible. Fortunately they had some of the island's best logistical brains in the room, including Ramón Gonzalez Cordero from Empire Gas, executives from Puma Energy, and Alberto from

Coca-Cola. If anyone knew about trucking needs, it was the head of Coca-Cola. There were officials from all the big government agencies, including the smart and quick-thinking U.S. Attorney in Puerto Rico, Rosa Emilia Rodriguez, and the island's secretary of state, Luis Rivera Marín. They had three problems to solve: more tankers to distribute the gasoline to the gas stations, more electricity for the gas pumps, and more security at the gas stations to deal with the long lines. They needed around a thousand security personnel to protect the gas trucks and the stations, and the National Guard offered up seven hundred. There were stories of people going to the gas stations with guns, but like all these stories, nobody had seen any trouble or actual guns. Within forty-five minutes the group had figured out a plan and the discussion was effectively over.

We moved on to another meeting about our real focus: food relief. There the contrast could not have been greater. Fuel for cars and trucks was a priority that attracted the best brains in business and government. But fuel for people seemed like a less urgent priority, as the meeting made clear. There was plenty of talk but not a lot of action. There were even people who wanted to take their photos with me and to stream interviews on Facebook Live. Thank goodness the Internet connection was too patchy to comply. I wasn't interested in the publicity; I wanted to see food relief in action. After an hour of listening to empty words, I grew frustrated and walked out.

We drove to the San Juan Coliseum, the city's biggest indoor arena, which was normally used for concerts but had been transformed into another distribution center. I knew the arena had a big kitchen, but my experience in Houston was that it wasn't easy to activate such kitchens, even when the public emergency was clear. We were told that the Coliseum, known as El Choli, was temporarily under the control of the first lady of Puerto Rico, the governor's wife. The arena was struggling: there was no power supply other than a few generators. But the kitchen was ideal for what we needed, and I had to figure out how to open it.

I met with Leila Santiago from the first lady's office, but the news wasn't good: we couldn't use the Coliseum kitchen because it was being used to feed the 150 people manning the distribution center there. This was a kitchen that could feed tens of thousands of Puerto Ricans, but it was only helping 150 people. There was some talk about the kitchen operators changing contracts and being shut down during the transition. Whatever the reason, I couldn't believe the lack of urgency and understanding. Surely someone understood how important food was? Kitchens were the biggest asset on the island, and the need to fire them up was obvious. This wasn't the result of bad intentions: people wanted to help, but they had no experience. Yet there's a world of difference between wanting to do good and knowing how to make it happen.

Ramón Leal, from the restaurant association, promised to find me another kitchen, at some former government offices. Driving from the Coliseum to view it proved difficult: our route was entirely blocked by a tree that laid across the main road. When we finally arrived, we found the kitchen was a disaster. It was a small kitchen for a café, if I'm being generous. If I'm being honest, I'd say it was the back room of a basement underneath a kitchen. It had no power and no generator. I have more cooking firepower in the garage of my house than they had in there. Two inches of water covered the floor because everything had melted out of the freezers. It was disgusting and would have taken a week just to clean up.

But this location did unlock one piece of intelligence for us: in a side room we found a huge supply of water bottles. We had been told there was not enough water on the island, but clearly there were supplies stashed away. Our challenge was to find those supplies, alongside a large working kitchen.

It was only our first day on the ground and I was already

frustrated. I felt like people weren't taking the food crisis seriously or addressing it with any real sense of urgency. I was worried that Puerto Rico would just become another Houston: a natural disaster compounded by man-made politics.

"Fuck it," I said. "Let's go to José Enrique to drink a rum sour." I was anxious to be with a chef, in a restaurant, where people were dedicated to food and cooking. José Enrique had promised me a plate of bacalaítos, salt cod fritters, with mojo sauce and my mouth was watering already.

The sun was close to setting by the time we arrived at my favorite restaurant in San Juan. There was no power in the historic Santurce district, where José Enrique's pink restaurant stands in what is normally a partying neighborhood of bars and restaurants tucked around a colonial-style market. La Placita, the market square, was quiet in the darkness, save for one giant advertising board that burned bright with its own power, promoting a concert that could never take place. I thought whoever was in charge of that illuminated billboard should be in charge of the whole electric grid on the island.

José Enrique's small generator was working as hard as it could, but we needed some extra light so we used the headlamps of our Jeep and some solar lamps. "Bienvenido," he welcomed me, offering a big hug and a bigger smile. Over rum sours, José Enrique told me how Santurce was struggling through the crisis and how popular his sancocho was. The lines for the stew were huge and the restaurant ran out of soup early. More and more people were showing up as the word spread. He brought me a plate of some leftovers, and it was delicious. The generator ran out of diesel, the solar lamps faded, and we drained our phones by using their flashlights. It was dark in San Juan, but that plate of soup filled me with love and hope.

This restaurant is a happy place for me. It's where my daughters love to eat when we come to San Juan, and we ate here often when I was opening my restaurant at Dorado Beach. José Enrique is a great chef and his family is as big-hearted as he is. He can look serious with his close-shaved head and beard, but his big smile gives him away. In these crisis situations, you need to find your fort, the home base that is your place of strength. I knew José Enrique's restaurant would be our fort.

José Enrique's plan was maybe to have a fund-raiser, with music. "People could have a little fun because it's so overwhelming," he said. But he also knew that he had a bigger role to play. "You need a cook who can feed you," he said. "It's kind of on us. Nobody else can do it."

José Enrique's problem was that he was doing everything from scratch. He and his cooks were making fresh *sancocho* every day, and the printout they gave me of their ingredients was so long and detailed, it was like reading Morse code. When you cook with the urgency of now, you have to cook quickly.

"Let's do more," I told him. José Enrique told me he had no more ingredients to scale up. They were cooking everything they had, because with no power and no refrigeration, the generator couldn't keep up and the supplies were sure to go bad. His cooks would only briefly open the fridge door when they absolutely needed to. Without much diesel, they could only run the generator for a few hours a day.

As we talked, our plan started to take shape. "Let's start right here," I said. We would grow José Enrique's operations as quickly as possible by using his kitchen, serving *sancocho* outside, and making sandwiches in the forty-five-seat dining room, where the customers would normally eat. I knew from my time in Haiti and Houston that sandwiches were a quick and effective way to feed people: plenty of calories in a meal that was easy to store and transport.

There was supposed to be a nighttime curfew in San Juan, as well as an order banning alcohol, but there we were, planning how to feed the people over cocktails, at night in the middle of the city. We decided to call ourselves what we were: Chefs For Puerto Rico. The name, and the hashtag, said it all.

WE WERE JUST A COUPLE OF CHEFS, WHO KNOW HOW TO COOK, TRYING to feed the many. Across San Juan, and back home in Washington, people with far more resources and supposedly far more intelligence were only just getting going.

On the same day I struggled into San Juan on one of the first commercial flights, two Trump administration officials visited Puerto Rico for the first time since the storm. Among them were Brock Long, the head of the Federal Emergency Management Agency, and Tom Bossert, the president's Homeland Security adviser. White House Press Secretary Sarah Sanders told reporters, "We've done unprecedented movement in terms of federal funding to provide for the people of Puerto Rico and others that have been impacted by these storms. We'll continue to do so and continue to do everything that we possibly can under the federal government to provide assistance."

That was a fantasy. Brock and Bossert returned to D.C. the same day. According to the Pentagon, 2,600 defense department employees were on the ground across the Caribbean, including Puerto Rico and the U.S. Virgin Islands.[6] On any normal day, there are many times that number of military personnel on the ground, between the U.S. Army's Fort Buchanan personnel and Puerto Rico's air and army National Guard.

The Pentagon had deployed the USS *Kearsarge,* an amphibious assault ship, and its group of sister vessels ahead of Maria to be ready to deliver essential supplies immediately after the storm passed. By the time we arrived, five days after landfall, they had airlifted just 22,000 pounds of supplies to Puerto Rico and the Virgin Islands. That's the equivalent of about thirty thousand bottles of water for a tropical island of 3.4 million people.

To put that into context, within two days of the catastrophic earthquake in nearby Haiti, in 2008, some eight thousand American troops were en route to deliver aid. Within 2 weeks, 33 ships and 22,000 troops arrived.

The need for leadership and swift action was no secret. "We need to prevent a humanitarian crisis occurring in America," Governor Ricardo Rosselló told CNN that day, warning that there would be a "massive exodus" of Puerto Ricans to the mainland if the island failed to recover. "Puerto Rico is part of the United States. We need to take swift action."[7]

Back in the White House that evening, American leadership took the form of a few tweets from President Trump, his first since Maria had devastated the island. "Texas & Florida are doing great but Puerto Rico, which was already suffering from broken infrastructure & massive debt, is in deep trouble," he shared after dinner with conservative members of Congress. "Its old electrical grid, which was in terrible shape, was devastated. Much of the Island was destroyed, with billions of dollars owed to Wall Street and the banks which, sadly, must be dealt with. Food, water and medical are top priorities—and doing well."[8]

According to a Trump official, the tweet was in response to the coverage of Puerto Rico he had watched on TV, not because of any meetings about the disaster that he had held that day. There hadn't been *any* such meetings involving the president of the United States. At dinner that night, Trump did make some brief comments about the tragedy in Puerto Rico but spent most of his time attacking Senator John McCain for voting against his efforts to undo Obamacare.[9]

Trump was right about the historical problems of the island in terms of finance and infrastructure. He was right about the destruction of the electrical grid. But it was not at all clear what he was doing about any of those challenges. And there was no earthly way anyone could honestly claim that the island was "doing well" with food, water and medical supplies. That much was obvious to me after one day in Puerto Rico.

NATE AND I DIDN'T HAVE AN ARMY. WE DIDN'T EVEN HAVE SATELLITE phones. But we did have a couple of hotel rooms booked at the AC Hotel near Santurce, and the Hyatt near the convention center, thanks to my board member Javier Garcia and my Puerto Rican friend Federico Stubbe. Sometimes you need to book extra rooms because there are unexpected guests to help or because the bookings mysteriously fall through. It was now 10:00 p.m. and I wanted to check out the AC first because it was originally part of a Spanish chain that I liked. But with the heightened security and the curfew, the driveway was blocked off when we arrived.

"Are you staying here?" asked the security guard.

"No. I'm meeting somebody," I replied, not wanting to explain why we had rooms in two hotels.

"We're closed," he snapped back.

It took far too long to convince him we had a reservation. The atmosphere of fear was widespread, and it may have been irrational, but it was no less real. If anything, the lack of accurate information just heightened the fear.

We finally made it to check-in, exhausted but inspired, after our first day on the island.

"We have no rooms," said the man at the check-in desk. "No rooms at all."

Nate urged me to head to the Hyatt, where we had another reservation for two rooms. But I wasn't about to leave. When people tell me something can't be done, that makes me all the more determined to get it done. Even when it comes to hotel rooms.

My friend Bernardo Medina, a media and communications expert, was staying at the AC and met us in the lobby. He tried to call the general manager, who was staying at the hotel, but couldn't reach him. After half an hour of back and forth, the staff at the check-in desk insisted that we leave. The security guys started making their move when the general manager appeared. He apologized for the delay; he'd been taking a shower.

I explained patiently that I knew the man who built the hotel: Antonio Catalán Diaz, the founder of this hotel chain and also the NH Hotel Group. Antonio is a Spanish entrepreneur and we'd been talking for years about doing business together. This seemed as good a time as any to drop his name into the conversation. It worked. Somehow the manager found us two empty rooms, where we gratefully rested our weary bodies. Nate and I had no idea that those rooms would become our home for weeks on end, and how our bones would ache for those beds every day.

WE WOKE UP EARLY ON OUR FIRST FULL DAY IN PUERTO RICO WITH ONE mission in mind: to get our hands on lots of food. I knew from my restaurant operation on the island that the biggest food supplier was José Santiago, so we drove our Jeep to his warehouse, twenty minutes south of San Juan. I was worried about how the business had survived the hurricane and whether they had electricity. Perhaps they might not be open because the employees were still taking care of their families.

On that car ride we saw for the first time with our own eyes the extraordinary gas lines. There were hundreds of cars parked, people waiting for a precious gallon of fuel for ten hours at a time. You could only buy $20 of gasoline at once, so people were lining up every day or two, and even sleeping overnight to stay in line. It felt like we were looking at an entire economy that had broken down by the side of the road. And I couldn't help but wonder: If people were waiting that long for energy for their cars, what kind of energy were they getting for their bodies? It gave me flashbacks to Haiti, where I saw similar scenes, and even more instability caused by the lack of fuel.

At José Santiago's headquarters there was another long line

of cars, as people waited patiently for their chance to enter the food distribution center. I saw some movement by the door, and realized the facility was operational but overwhelmed. Nate and I couldn't wait for the line to move, so we jumped ahead. Once inside, I introduced myself to the grandson of the Spanish founder of the business, who shares Santiago's name and serves as chief financial officer. As we were walking through the back offices, I noticed a picture on the wall of my historic navy ship, the *Juan Sebastián de Elcano*. I could recognize its four masts from a mile away, and it didn't take much to trigger the memories of my year sailing the world on that majestic tall ship. As we talked about the ship, José told me of his family roots in the same northern region where I was born: Asturias, famous for its green pastures, its dairy and its apple cider. It felt like he and I were family, and I asked for a line of credit right there.

"I am from Asturias. You are from Asturias," I told him. "I won't let you down. We will pay you for the food. You don't need to worry."

We were handed a full catalog of supplies, printed on one long extended run of dot-matrix paper, which folded up to be three or four inches thick. It represented a world of food on an island struggling with hunger. Then, on the spot, José and I shook hands to agree to a $50,000 line of credit. Apart from the paperwork, he had one other condition: we couldn't tell people publicly where we got the food. No photos for social media, and no talking to the press. He too was scared of the rumors of lawlessness and hunger, and feared a mob descending on the warehouse to loot its supplies. "We don't want the word to get out," José told me. I would have done the same in his situation.

Before we arrived, we thought we would just buy a few items for that day's lunch. Instead we walked around the place, twenty rows across and six shelves high, stocking up on the kinds of ingredients and quantities we'd need to scale up the operation at José Enrique. We brought huge trays of bread for sandwiches, and plenty of ready-made *sofrito* for the soup. We filled the Jeep from floor to roof, and happily spent around $5,000 on our first purchase. That would end up being a relatively small bill. There was so much food, I couldn't see through the windows on the right side and needed Nate to tell me if the roads were clear.

We drove over to the restaurant very slowly. The last thing I wanted was an accident, creating more trouble for the island's hospitals. I had asked José Enrique to gather together the best people who could help put a food relief operation into action. We were late getting there, but our Jeep was full of ingredients for meals, so people seemed happy to see us. Seated around the sides of José Enrique's dining room were some of the smartest restaurant people in San Juan: Wilo Benet, the chef whose Pikayo restaurant helped reimagine Puerto Rican food; Ricardo Rivera Badía of El Churry; and Manolo Martinez of Paella y Algo Más. Our organizer would be Ginny Piñero, a former lawyer, who knew Manolo's son in Washington, D.C. She had no idea what was in store, and nor did any of the chefs.

I walked in and started mapping out a plan, assigning tasks to everyone there. I propped up a flip chart under a painting of a giant green flower, while the team sat at some empty dining tables pitching in their ideas. I wanted them to feel some sense of ownership of this plan: I didn't want to impose it on them, not least because I planned to leave by the next weekend. This was something they needed to own, and they stepped up immediately.

At the top of the white sheet, I wrote in huge purple letters our biggest challenge: ENERGIA. Gasoline, natural gas and diesel: we needed them all. We couldn't do anything without them. We assigned this challenge to Piñero, who looked like she was in shock.

Next up was the energy we needed as people: ALIMENTOS.

We needed dry goods and fresh goods, and especially water. We assigned this to Ricardo, whose experience inside the food business in Puerto Rico was second to none.

Those two items filled up the left side of the flip chart paper. On the right, I wrote our next most important need: VOLUNTARIADO. We desperately needed volunteers: cooks, cleaners, people who could help prep food and buy ingredients. We needed coordinators of the volunteers, and we needed people to help with distribution. I assigned this task to José Enrique's sister, Karla.

That left our last, but connected, necessity: COMUNICACIÓN. How to get the word out about our operation? On an island like Puerto Rico, there was a powerful combination of old and new media. People relied on the radio, especially given how much time they were now spending in their cars. There wasn't enough electricity to turn on their TVs. But they were also, like in every other part of the United States, busy with social media on their phones, if they could find a signal. We needed to be in both places. I handed that assignment to my friend Nate and Yareli Manning, one of a group of food truck owners who showed up from the start.

The team sat there, focused on the ambitious plan in front of us. It was barely twenty-four hours since Nate and I had landed in Puerto Rico, with nothing more than a modest pile of cash and a desire to feed the hungry, yet we were already mobilizing a much bigger operation than we'd expected, and detailing the steps we would need to take to get even bigger, very quickly. We weren't trying to feed the island; that would have been an overwhelming challenge. We didn't want to make anyone feel anxious because they had more than enough problems already, looking after their family, their home and their business. We were just trying to double the meals being prepared, and double them again. We needed to grow and grow until we were feeding more people than we could imagine at that point. It was just like opening a new restaurant: we could reach our maximum capacity over time, scaling steadily but surely. That day-by-day approach to exponential growth seemed much more realistic than shooting for the moon.

Ginny Piñero volunteered to be the point person for taking food orders, and we started receiving orders that day. She seemed to know everyone on the island and was well connected by social media to a wide network of Puerto Ricans. We took the most important organizational step of all by setting up a WhatsApp group that night, using our brand-new name: Chefs For Puerto Rico. That chat group would become our central intelligence agency for the next several weeks, a constant buzz of information, questions, demands and good humor. It was our team spirit and lifeline, tapped out on phones from the most disconnected parts of the island and the darkest corners of kitchens. It also worked with the faintest of cell phone signals, while phone calls and emails dropped in and out unpredictably.

Sourcing supplies on a hurricane-torn island was never going to be easy. But Ricardo knew the inside workings of the restaurant networks in Puerto Rico, and could find people and products that nobody else knew existed. Ricardo is a big bear of a man, but with a quiet, no-nonsense style. In many ways he was typical of the Puerto Ricans who formed the backbone of our food relief on the island. He was a franchise restaurant consultant of many years, helping others to expand their businesses across the island. More recently he'd started his own El Churry franchise, expanding a successful food truck operation into full restaurants. He survived the hurricane at home in Caguas, with his wife, Luz, and two dogs, taking shelter in the bedroom as they heard their trees getting wrenched out of the ground and thrown into their neighbors' house. The next day they emerged, along with the neighbors, to assess the damage and remove the branches that had fallen on their roof. For three days they were stuck in their neighborhood because the roads were locked with trees and electric cables. Ricardo was one of the lucky ones: his Internet connection survived the storm. So when his wife shared a Facebook post from the Puerto Rican governor's office in Washington, D.C., he paid attention. They were looking for food trucks to head to the airport, where there were stranded tourists with no food, and Ricardo said he would help out.

We immediately got to work with a new mobile kitchen parked outside José Enrique's stricken restaurant. We unpacked the Jeep and set about our first official day of operations. We knocked out five hundred sandwiches and two thousand hot meals, and considered it a triumph to feed so many people from a broken restaurant on an island struggling to get back on its feet. We thought those were big numbers, but we knew we could do so much more. We just had no idea how much more was in store.

That afternoon, I ladled *sancocho* out of a giant stock pot on the front steps of José Enrique's beautifully pink restaurant. There's normally no nameplate or sign outside the restaurant because it's so famous. But now, behind me, was a simple sign handwritten by José Enrique, *Hay Sancocho*, telling everyone in the neighborhood that we had soup for them. The streets were full of people eating, talking, connecting. It felt like, in this small corner of San Juan, we were bringing the island slowly back to life, one ladle at a time.

I had another chef friend across Santurce who also wanted to help but was out of the country, visiting Morocco when the hurricane ripped through. José Santaella is one of the very best chefs in Puerto Rico, who like José Enrique is reinventing the island's culinary traditions. When I reached out to him for help, he said, "Go into my kitchen. Go into the freezer, find your food and use it." The freezer had lost power, and was filled with a huge amount of the highest-quality food.

That afternoon we returned to our food suppliers, José Santiago, to pick up ingredients for the next day. We spent more than five times what we shelled out the day before: $26,000. We were growing, and growing quickly.

We weren't the only ones to step up our operations that day. Back in Washington, D.C., President Trump held his first meeting to coordinate the response in Puerto Rico, in the Situation Room inside the West Wing basement. The reports from the island were so bad that later in the day the White House added a Cabinet-level meeting on the disaster. The Pentagon ramped up its efforts, sending the hospital ship USNS *Comfort,* as well as several transport planes full of heavy-duty trucks to help with the recovery. Perhaps the most important resources were on the planes carrying equipment to help restore operations at the airport and establish satellite communications.[10] Both U.S. senators for Florida—Republican Marco Rubio and Democrat Bill Nelson—wrote to the president asking him to "leverage all available resources" to respond to the hurricane, especially the U.S. military. "Our brave men and women in uniform are well equipped, trained and tempered to handle the dire situation in Puerto Rico and the U.S. Virgin Islands," they wrote.[11] Their letter was proof—if anyone needed it—that the military was not, in fact, deploying all its resources to respond to this overwhelming disaster on U.S. soil.

It was now six days after Maria made landfall, and Trump called the Puerto Rico governor for just the second time since the hurricane.[12]

The situation on the ground remained dire. According to the Pentagon, 44 percent of the island was without drinking water in the tropical heat. Just eleven of the island's sixty-nine hospitals had any fuel or power. As for the power grid, it was almost nonexistent: 80 percent of the transmission network and fully 100 percent of the distribution network was damaged. The island was flatlining.[13]

That night we picked up some avocados for ourselves on the way back to the hotel. One single restaurant in the neighborhood had kept its lights on: a beer and burger joint across the street, called The Place. The manager of the restaurant had been smart enough to do a special deal with Sam's Club before the hurricane swept in, because the supermarket was worried its ground beef would go bad after what they expected would be a long power blackout. So they sold their vast supply of beef at a steep discount to The Place. Everyone on the island was desperate for cooked meat at this point, so the lines for their burgers were insanely long.

I took my avocados to the hotel kitchen on the penthouse floor, but it was closed and the cook tried to kick me out. I would do the same if a stranger came into my kitchen. I asked her if I could make myself a salad, but she was worried I would cut myself with the knife. I always respect the kitchens of other chefs, but I assured her I had a bit of experience with kitchen knives. We ate my avocados in the bar area, and served many more to the relief workers and hotel staff who were alongside us. The best friend-ships often start with a simple plate of food, accompanied by a drink or two, like my favorite Caribbean cocktail: the rum sour, my sweet taste of heaven back in Haiti.

In the bar were several federal law enforcement officers, whose agency initials I didn't recognize: HSI. Homeland Security Investi-gations was a relatively new security force merging several groups when the Homeland Security department was created after the 9/11 attacks. HSI was part of Immigration and Customs Enforce-ment, known for deporting undocumented immigrants. In contrast, HSI normally chased after human smugglers, child traffickers, and drugs and arms dealers. But here they were checking on the safety and security of government officials in the middle of the disaster. When everything else breaks down, HSI's special agents travel into the most difficult places to make sure that the federal government's employees and families are safe and ready to get back to work.

They looked ready for a war zone. Even in the penthouse bar of a luxury hotel, they were wearing flak jackets and carrying sidearms. Maybe I've watched too many movies, but I have al-ways loved the idea of special agents. And in this crisis, the sight of a well-trained force in uniform looked to me like a welcome relief after seeing so little organized response to the hurricane on the ground. I started talking to the agents to understand their mission and see if there was some way we could work together.

My wife had asked me before I left for Puerto Rico if I could track down a missing eighty-year-old relative of one of her best friends. He was somewhere on the western side of the island, in Añasco, where there were no communications. I was trying to drive there myself, but everyone told me I was crazy to consider such a trip. The HSI agents said this was the kind of work they were doing across the island, as they investigated the state of law and order on the island. They took the relative's name and address—and the phone number, in case it worked—and promised to get one of their patrols to help, as they continued to drive across Puerto Rico.

These disaster zones create some strange connections as talented and creative people gather to restart their lives. That evening, I got stopped in the hotel lobby by a chef who recognized me. He was gathering supplies to help the recovery effort in some part of the Virgin Islands. I later figured out he was working for Google's Larry Page, who owns an entire island called Eustatia. His nearest neighbor is Virgin's Richard Branson, who lives on Necker Island. Those places, blessed with such wealthy owners, were hit even harder than Puerto Rico, with no easy way to reestablish their supply chains.

THE NEXT DAY WE MET EARLY AT JOSÉ ENRIQUE'S RESTAURANT TO MOVE the operation into high gear. My team was already searching for fuel and food supplies, and we were having more success with the food than anything else. It was clear that access to food was not the challenge on Puerto Rico, despite the conventional wisdom that the island was too remote or too stricken to feed itself. The real issue was distribution, and communication was at the heart of that. Ricardo and Ginny were already reaching out to the media, especially radio shows, to tell people what we were doing. But I was cautious. As much as I love the media, it was important to be realistic.

"Be careful," I told Ginny. "We don't want to say too much and create false expectations. Let's go day by day in terms of the number of people we can serve."

Nate and Ricardo drove two Jeeps to our food supplier to fill up with the day's ingredients, as well as a thousand aluminum serving dishes, but they called me up sounding frustrated.

"They are shutting us down," Ricardo said.

"Why?" I asked.

"Because we've spent the $50,000 line of credit already."

I called up Santiago to give him hell. "I want thirty-one days," I said. "I am a businessman." He backed down because he knew we would become his biggest customer in no time. And because we were all—Santiago included—helping to feed the people of Puerto Rico. It's an old saying but it's still true: Where there's money, there will always be food.

I started looking for the greatest areas of need. Through a Puerto Rican friend living in Paris, Daya Fernandez, I was given the number of her uncle, who was a doctor at the biggest medical center on the island: the University of Puerto Rico's hospital, known as Carolina.

The crisis at the island's hospitals was as severe as you could imagine. At the Carolina, as at the main Centro Medico hospital, the hurricane had inflicted enough roof damage to flood the wards below. The generators were old and unreliable, meaning what little power they had would cut in and out. Air-conditioning, ventilators, operating theaters were all limping along. The result, among sick people in tropical heat, was predictable. If they were injured in the floods, or already sick before the hurricane, their chances of survival were now much, much lower. At the Centro

Medico, one of the hospital's three generators was down. When they asked the federal government for a replacement, they were turned down because the Army Corps of Engineers said they weren't considered "a critical need." They already had two generators, after all.[14] Local mayors were telling the media that people were dying because of the lack of power, and warning that the hospitals were at capacity. Some doctors were just telling people to get out of Puerto Rico. "If you are sick in Puerto Rico," said one surgeon at the cardiovascular center in San Juan, "the best thing is to get on a plane and abandon the island."[15] At this point, the official death count of just 16 people seemed like wishful thinking. There were credible reports that the hospital morgues were at capacity, including at the hospitals that were closed (which was 70 percent of the island's medical centers), and at the hospitals that were cut off from normal communications.[16]

The stress on the hospital staff, as well as the patients, was huge. How could you keep a hospital clean in these conditions, when the temperature would spike in the humid heat? With all the struggles they faced—the hurricane damage and lack of power—the staff had no time or resources to feed themselves. It was hard to believe that no one was thinking about taking care of the essential people caring for the Americans in greatest need. They could hardly disappear for a lunch hour to find something to eat in the neighborhood. Even if they could, there were no restaurants open or food trucks on the street.

"We have 400 employees and they haven't eaten here for the last week," Dr. Carlos Fernandez Sifre told me. "They are working overtime and nobody has money or electricity. The ATMs don't work and the markets are empty."

Immediately I realized we needed to identify the most important groups to feed. They needed to know we were ready to help. I promised to send the hospital two hundred meals that afternoon and made sure we told everyone on social media what we were doing.

Our press and social media operations were not some vanity project. It was vital to tell the people of Puerto Rico that we could help, and that the world cared. And you never knew how they might hear the message. After all, I only heard about the hospital via a friend in Paris. The local media—Univision, Telemundo, and local radio stations—started to come to us and we were happy to provide information. We desperately needed to get the word out. I handed my phone to Nate with a simple order: "Tweet." I was too busy growing and running the food operations to focus on social media through the day, so Nate became our eyes and ears on the world. He frowned and looked worried, but he knew what I wanted to say and when I wanted to say it.

That morning was the first time we had our cooking lines all fired up. Manolo was making a huge quantity of rice and chicken in giant paella dishes outside. Inside we were cranking out hundreds of sandwiches with cheese and ham. And our favorite, *sancocho,* was the mainstay of the central kitchen itself. The problem we now faced was that we were quickly running out of the basic tools of the job: big stockpots, knives and cutting boards, and aluminum trays to transport and serve the food.

We used most of our trays to take the meals to the hospital, filling up the back of the Jeep with hot food and bottles of water, as well as fresh grapes and apples, and a pot full of *sancocho.* Finding the hospital was a challenge in itself, with all the street closures and fallen poles. When we arrived, the staff was so grateful they asked us if we needed anything in return. We assured them we would be back again tomorrow, and for as long as they needed feeding. We shook hands to say they could rely on us.

While we were at the hospital, I caught a glimpse of the depths of the disaster after the hurricane had passed through. I saw one dead body on my brief visit, and three more on later hospital visits, which was about as many as I had seen in my whole life before Maria. My parents were both nurses, and I thought I was used to seeing the worst in intensive care and the emergency room. But this seemed wholly out of line with the official story, that the death toll was only a dozen or so Puerto Ricans. If the numbers of victims were so low, how could I have seen one-fourth of all the corpses on my own trips to the island's hospitals? The numbers just didn't add up. It was already clear to me that this was a deadly serious humanitarian crisis. It was also an untold disaster, hidden from view and lied about by our public officials. My mission was to help my fellow American citizens, and to tell their story to a world that was living in the dark.

Back at the restaurant, our cooking operation was busy enough to fill up our first round of food trucks, which were our secret weapons. Instead of asking people to come to us, the food trucks could find the communities in greatest need: the people who had no gas or cars to drive to us, or were simply too old, too sick or too busy to travel. Nobody thinks about food trucks because they are still considered an informal or unserious part of the food distribution business. To me, they were ideal for our mission: a moving kitchen, perfect for storing and distributing food across large distances. Further, the trucks were simply available: after the storm they weren't doing their normal runs, because their usual customers were no longer able to work and nobody was going to stand on the street, waiting for a tasty lunch at a great price.

We stocked up one food truck, which normally sold Chinese food under the name Yummy Dumplings, with 250 meals to deliver in the El Gandul quarter of Santurce. The neighborhood is poor and known for its social problems, but its name—like so much else on the island—comes from a forgotten history of its food. The *gandul* is the pigeon pea, which used to grow there in large quantities. Now there were many reports that the entire neighborhood lacked food and water.

An hour later we were serving *sancocho* at the restaurant, to a street that seemed to fill with more people every day. Among the people who just showed up: the inhabitants of the municipality of Caguas, a town in the mountains about thirty minutes south of San Juan. They said they needed 650 meals for a home for the elderly, and we were happy to help with our first giant paella pans, filled with rice and chicken. The news of our food relief was traveling farther, and with it, so was the food.

As we expanded, we needed more capacity to support the rapidly growing kitchen. For more space, we rented the car park opposite the restaurant, paying the owner as if the place was permanently full of cars. With help from the restaurant association, the island's secretary of agriculture, and the first lady's office, we took delivery of a giant refrigerator truck that we could park outside the restaurant. José Enrique's walk-in fridge had too little power and space for what we needed, and this huge trailer was a vital foundation for what we planned to build. The challenge was getting our hands on enough fuel, but José said we didn't need to worry. We started feeding the local police station, and they gave us diesel in return. For now, we were doing fine for fuel.

That evening we drove to Univision radio to talk to the popular host Jay Fonseca. Even in times of crisis, you expect to find a handful of working organizations: a government headquarters, perhaps a supermarket or two, and the news media. But we couldn't believe what we saw at Univision: a giant satellite dish lying crumpled in the middle of the road like a toy thrown from a passing car. There was no security in sight and no obvious way into the building. It took me twenty minutes to find a side door that was open.

Radio proved to be our best friend. Ginny had been interviewed on the radio that day and was asked a simple question: How could people tell us they needed food? We didn't have a great answer to that, so Ginny just answered honestly. "Call me," she said, handing out her number on the air. After that, her phone never stopped ringing with orders.

That day we prepared 4,000 meals. We doubled our sandwiches from the day before, from 500 to 1,000, and we grew even faster with the hot meals, going from 2,000 to 3,000. It was a big step up, and it was a sharp contrast to what we saw from the officials in government, who had so many more resources at their disposal.

IT WAS EIGHT DAYS AFTER MARIA MADE LANDFALL, AND THREE DAYS after I arrived in Puerto Rico. It felt to me like we were still just discovering the true impact of the hurricane: who needed help and where they were. So many places, like the hospitals, were in desperate need. Our goal was to cook 5,000 meals but the system was straining at the seams. Our orders and deliveries were patchy. If we had enough bread, we didn't have enough cheese for the sandwiches. If we had enough rice, we didn't have enough aluminum trays to deliver the hot meals. There was lots of activity but it was hot, sweaty and frustrating.

Since we were already at capacity at José Enrique's restaurant, there was only one solution: open more kitchens. That's what we did at Mesa 364, a restaurant across San Juan run by chef Enrique Piñeiro, an intimate, high-end place at the forefront of contemporary Puerto Rican cooking. He partnered with a volunteer group called Mano a Mano, which brought families together by establishing contact with missing relatives. The group was set up by Lulu Puras, who also knew my Parisian friend Daya. Lulu normally ran an interior design store, but realized quickly that she had a new mission. "People are suffering too much," she said. "How are they going to order a lamp or a rug?" So she closed her store and told her customers it was an emergency. As they traveled across the island, Mano a Mano could deliver food and water, at the same time as they reassured people about their loved ones. We sent them supplies of chicken and rice to help meet our targets, which now included both the major hospitals—the Centro Medico and the Carolina—as well as several homes for the elderly, known as *egidas*.

That afternoon, we met with our first media from the mainland: the intrepid Bill Weir from CNN. Bill had ventured farther than anyone across Puerto Rico, renting a plane to travel to the forgotten island of Vieques, where there was little police presence. The airport was strewn with planes ripped apart by the hurricane, and the roads were barely passable with cables dangling overhead. In the center of the small island's main town, he showed Puerto Ricans in tears as they used a satellite phone to ask relatives to send food and water. Former politicians told him they were living in a war zone and pleaded for marshal law to be imposed so they could move around at dark, rather than live in a no-go area.

We messaged Bill through Twitter to thank him for his reporting and told him we would love to help the people of Vieques. He responded by asking if he could come see our operation in Santurce. Given our operational problems that day, I wasn't sure this was a good idea. We needed to show the world a kitchen in full swing. But soon after we started serving our daily *sancocho*, he showed up anyway. The lines for the soup were huge that day; word was spreading about what we were doing. I walked Bill through our mobile kitchen and refrigerator trailer, and talked him through the giant flip-chart sheets that showed our daily orders and output of meals. That evening we started planning our first trip to Vieques to see for ourselves what Bill had shown briefly on CNN.

He wasn't our only surprise visitor. Soon after he left, the Salvation Army stopped by unannounced. It was getting dark, and starting to rain, and our cooking was almost shut down for the day. "We're looking for a kitchen run by chefs," said Captain Don Sanderson.

"Well, this is your lucky day," I told Sanderson, as I smoked one of my Arturo Fuente cigars.

He wanted to know if we had any food we could give them to deliver to a shelter full of elderly people.

"How many people?" I asked.

"Two hundred," he said.

"Do you want sandwiches?"

"Yes!"

"What about fruit?"

"You've got fruit? I'll take it," he said, smiling.

"Do you need water?"

"Yes, sir. God bless you," said Sanderson.

It was only then that we realized we were a long way from being a small and scrappy start-up. Sanderson mentioned something about a FEMA meeting the next day, run by a group focused on what was called "mass care," suggesting I come along. It sounded like a foreign language to me. But if the Salvation Army—one of the very biggest charities in the world with annual revenues of $3.7 billion—was knocking on our door, what did that say about food relief in Puerto Rico? It was stunning to think that there was nowhere else for them to go. Maybe we had no reason to feel inferior in Bill Weir's presence about the size of our operation.

The dry law that banned alcohol was lifted in Santurce, and the small bars around José Enrique's restaurant started to reopen. It was great seeing the neighborhood I loved come back to life. But the extra visitors, combined with our lines for *sancocho,* were creating a scene. Our operations were filling the car park opposite the restaurant, and the historic narrow streets were straining under the load. Big delivery trucks couldn't turn the sharp corners and needed to back up for blocks. There were paella pans bubbling away outside, while the restaurant itself had become a glorified storage facility. We were clearly outgrowing the Santurce neighborhood. We needed to think about the next phase if we were going to feed the island as it desperately needed to be fed.

CHAPTER 2

FEED THE WORLD

SEVEN YEARS BEFORE MARIA, I GOT MY FIRST TASTE OF FOOD RELIEF ON another Caribbean island.

I didn't know what I would find on my first visit to Haiti. The earthquake of January 2010 flattened much of Port-au-Prince, including the presidential palace and national assembly building, as well as a quarter of a million homes. It stole as many as 158,000 lives, and prompted international aid of more than $5 billion. The place was overflowing with aid workers, doctors and nurses. What could a chef really accomplish there?

I visited with a correspondent from El Mundo, Carlos Fresneda, along with a friend who had a brilliant idea: solar-powered ovens, giant metal parabolas that reflected the sun's rays to focus all that natural energy on a pressure cooker at their center. My friend Manolo Vílchez believed they could be the solution to the food crisis in disaster zones, and I thought he might be right. Not just about the solar ovens, but about the mission: maybe food was the solution we needed.

We landed in the Dominican Republican and crossed over the border to Haiti a few weeks after the earthquake, connecting with a Spanish nonprofit called CESAL. The scenes in Port-au-Prince were shocking and Haiti's effect on me was profound. I realized I needed to do much more than a single chef or a single visit could achieve. I would need an organization, modeled on my volunteering back home in Washington, D.C., for the extraordinary nonprofit called DC Central Kitchen. So I adapted the name to create World Central Kitchen, eight months after the earthquake, thinking it would become a model for what we could achieve through the transformational power of food. I ended up traveling to Haiti more than two dozen times, as we developed smart solutions to hunger in five different communities. One of my main goals was to educate islanders about clean cookstoves, using solar power and natural gas instead of the charcoal and wood they had long used. Normal cookstoves belch smoke into the homes, making the women and children there sick. Young girls spend hours gathering wood for the stoves instead of studying to break the cycle of poverty. By cutting down the forests, the islanders had hurt their own farms, as the rains washed away their precious topsoil. Mudslides destroy farms and homes, and pollute the local waters to the point where fish and coral reefs cannot survive. You can understand my passion: if you can give women control of their cooking, you can feed and heal an island.

My dream was to find a way to feed the many in ways that would help the local economy. We could create a network of chefs, like Doctors Without Borders, to help in a crisis. Rather than dumping food aid on an already struggling economy, we would source our supplies locally, wherever possible, and help put the farmers and suppliers back in business. Ultimately, we would develop viable food businesses—from farms to restaurants—that could help deliver local services to the people in greatest need.

But first, we needed to understand the people and what they expected of their food. It wasn't good enough to be a chef; I needed to learn, and a group of Haitian women were my teachers. On my first visit, I had been cooking a bean stew for hours on the solar stove, and I planned to serve it with some rice at a camp. These women looked at my beans and said thank you, but that wasn't the way they liked their beans. The beans needed another hour of cooking before I could puree them into a sauce they liked. My first reaction was frustration. But then it hit me: if this was your only meal of the day, and perhaps your only hot meal in several days, you too would want it prepared your way. So I strained the beans through part of a U.S.A.I.D. bag that used to be full of rice. In an emergency you need to get creative. At these times, you need to know when to lead and when to follow, when to teach and when to learn. These women taught me a lesson I could never learn in a classroom, or dream up in the office of an international aid group. A plate of food is not just a few ingredients cooked and served together. It is the story of who you are, the source of your pride, the foundation of your family and community. Cooking isn't just nourishing; it's empowering.

Haiti also taught me what people, even in the most desperate situations, think of military food that comes plastic-wrapped in bags that can be dropped by parachutes. Military MREs, or meals-ready-to-eat, can survive heat, cold and floods. They also taste like they can do only that: after eating three of them, you never want to see another. In Haiti I saw some kids playing soccer with an MRE. It was obvious an MRE could survive being kicked around for hours on end, but it would never represent real food to anyone.

Cooking food is one of the very few features of human life that makes us different from other animals. Some evolutionary biologists believe that our brains developed around 1.8 million years ago precisely because we found a new way to consume calories without using lots of energy to chew and digest them: by cooking meat.[1] Along the way, we evolved by eating cooked food, as our social customs revolved around cooking and eating together. We communicated around the food and fire, developed language and love, told stories of who we were and who we wanted to be. We may have changed greatly over the millennia, but in a disaster we revert to some very basic needs. Yes, we need water, food and shelter. But we also need our food to represent something more than food, if we are to rebuild our lives. Meals need to be cooked for our communities to come back together.

In Haiti, our ideas grew to long-term community-building through food, delivered through local partners. In 2014, we opened a bakery and a fish restaurant in Croix-des-Bouquets, on the grounds of a Partners in Health children's home, with the help of my friends Jean Marc and Verena de Matteis. The bakery, Boulanjri Beni, bakes bread for the children's home and sells more to the public. Two pastry chefs from my restaurants helped train five staff to bake the bread, and we donated all the equipment, from mixers to ovens. Because of the disabilities of some of the kids at the children's home, we built the bakery to be accessible to all. "Some of them will never leave," says Loune Viaud, executive director of Partners in Health in Haiti. "We *are* their family." The restaurant, Pwason Beni, opened soon after, and cooks fish from its own fish farm, which was created by the impressive foundation Operation Blessing. Pwason Beni started just serving the staff, but we soon opened it up to the whole community. The revenue from both businesses support the children's home, meaning the food serves the community in three ways: by feeding customers and children, employing staff and suppliers, and funding the children's home itself.

Beyond our local social enterprises, we knew we had to help the whole island. We converted more than a hundred school kitchens to clean cookstoves and trained more than seven hundred cooks in food safety and sanitation. We also opened a culinary school in Port-au-Prince in 2017 to train forty students a year to help grow the island's hospitality industry, after helping to create the curriculum three years earlier.

We took our social entrepreneurship to the Dominican Republic, with a honey business called 21 Women Honey, supporting a whole community of women and their families. Our support doubled the number of hives there and more than doubled the community's honey production. We traveled to Nicaragua, where we invested in coffee roasters and packaging equipment to stand up Smart Roast, and helped lift the income of coffee farmers by 400 percent. At two schools—one in Haiti, another in Zambia—we developed vegetable gardens, chicken coops and a bakery to help feed the children and sell the surplus for extra revenue. Soon we were operating in eight countries, including Brazil, Peru and Cambodia.

THE SOURCE OF MY INSPIRATION WAS BACK IN MY HOMETOWN. DC Central Kitchen is an amazing idea that has delivered incredible results. At its heart, the nonprofit founded by my friend and mentor Robert Egger is a giant recycling organization. It takes leftover food from hotels and restaurants, and turns those leftovers into meals for the homeless. The "Central Kitchen" name comes from the fact that it was founded in the basement of what was then D.C.'s central homeless shelter. The kitchen didn't just feed the homeless in its own shelter; it trained some of them to become cooks by feeding their own community, so they could later get hired by a restaurant or hotel.

Egger started out as a young nightclub manager who was frustrated by the traditional approach to the hungry and the homeless. He wanted to reinvent the very notion of a soup kitchen, creating a social enterprise that could fund itself. Coming from Washington, he believed his vision had something for people on both sides of the aisle. Democrats could support the community spirit of caring for those in need; Republicans could support the premise of self-help that lay at the heart of the work. Egger was also great at evangelizing about his model, helping to start more than sixty similar community kitchens across the country, as well as a food recycling program for college campuses.

Egger showed me the way. I joined as a volunteer but he quickly brought me on to the board of DC Central Kitchen. It was his example that led me to create an international version of his original idea: a network of chefs and kitchens that could serve the hungry across the world.

IN MANY WAYS, HAITI WAS THE BEST PLACE TO DEVELOP WORLD CEN-tral Kitchen because it was also the worst place for international aid. If you want to learn about the broken state of disaster relief, there is no better school than Haiti.

Haiti was broken well before the earthquake of 2010. During a century of French rule, one million slaves were forced onto the western half of Hispaniola to work on the sugar and coffee plantations. When the former slaves rose up in rebellion in 1791, the French and Americans retaliated. Napoleon's invasion was defeated but an American embargo was far more damaging. In 1915, the U.S. invaded and took control of the country's finances, after Haiti had suffered a century of debt to France for the land and slaves it had lost. After the Americans left in 1934, Haiti suffered decades of dictatorship and misrule. The farming economy was destroyed by cheap U.S. imports, subsidized by the American taxpayer. Crops, as well as forests, began to disappear. In the years before the earthquake, there were food riots, floods and hurricanes. If you wanted to change Haiti, you didn't work with the government; you worked through the aid workers, who were the new foreigners in charge.[2]

After the earthquake, the world was genuinely shocked by the scale of the human suffering and loss. Regular American citizens put their money where their hearts were: private donations to organizations like the Red Cross reached more than $1.4 billion. Rescue and aid efforts were uncoordinated, in the absence of a working government and amid fears of civil unrest. Reports of looting, including at a World Food Programme warehouse, turned out to be false. Rather, Haitians were often found in wrecked buildings either salvaging their own property or searching for food and water. Many people expected violence, but those expectations were based on movies or bad journalism, not experience. There was no lawlessness after the San Francisco earthquake and fire of 1906, just as there wasn't after the 9/11 attacks or the London bombings four years later. Yet the fear of chaos and panic clouds the judgment of aid workers, leading them to issue top-down orders that are out of touch with the reality on the ground.[3]

Haitians were all too familiar with the mistreatment and misinformation. They have been linked to HIV/AIDS ever since the Centers for Disease Control issued guidelines in 1983 that wrongly suggested they were carriers alongside intravenous drug users and homosexual men.[4] When cholera broke out on the island for the first time ever in Haiti, the assumption was that Haitians were the source. In fact, cholera arrived in Haiti with UN peacekeeping troops from Nepal.[5]

If the violence wasn't real, the hunger most definitely was. Among the widespread scenes of desperation were written signs pleading for food and medicine. But it was hard to know the scale of the food crisis: aid agencies could not reach the Haitian agency coordinating food security because its office was destroyed. The result was chaotic and inefficient. Huge numbers of water bottles were shipped in, at great cost. To this day, the island remains full of plastic waste that blocks drains and canals after heavy rains. Air drops of food and water were tried and abandoned as ineffective and unsafe. The direct distribution of food led to chaotic scenes and long lines for cooking oil and rice, manned by fearful soldiers who fired warning shots or pepper spray. Most Haitians believed at the time that food should have been distributed through church and community networks, by the islanders and for the islanders. Instead, the volume of free food kept local food vendors and suppliers out of business.[6]

What happened to all the money spent on aid in Haiti? Through the year after the earthquake, $2.43 billion was spent, and the vast bulk of that—93 percent—paid for UN and NGO staff and supplies. Fully $151 million of that money went missing and couldn't be traced at all. Of the U.S. government's spending, $1 billion went to contracts, but only $4.8 million of that was spent on Haitian suppliers. The Pentagon spent $465 million on its own operations, including $1 million a day for a supercarrier harbored in Port-au-Prince for 18 days. Individual examples of waste told the bigger story: for some reason, the navy spent $194,000 on photo and video equipment in Manhattan, and the coast guard spent $4,462 on a deep-fat fryer.[7] What were the results of all that manpower and resources? The U.S. military distributed 2.6 million bottles of water and 4.9 million meals over six months to a population of around 10 million people.[8]

Of the estimated $3 billion donated to NGOs, the American Red Cross raised $486 million on its own. With just two dozen staffers on the ground, it struggled to spend the money. Six months after the earthquake, it managed to sign contracts for (but not spend) less than a third of the money raised. Most aid groups refused to say how they spent the money they raised, or if they spent it at all. Doctors Without Borders was rare in asking people to donate to its general funds because it admitted it couldn't spend the 30 million euros it raised for Haiti.[9] Most of the government money was pledged but never arrived, and much of the money donated was never spent. The dollars that made it through to Haiti were wasted on international staff who had minimal impact on the actual disaster experienced by Haitians.

When anyone bothered to ask Haitians what they wanted from reconstruction, the answers were mostly about independence and self-help. "For us to be adults, we must be able to feed ourselves," said one Haitian, who took part in a focus group trying to find the right solutions. "If they really want to help us, they need to invest in agriculture."[10]

THE QUESTION OF HOW AND EVEN WHETHER TO HELP PEOPLE IN A DISASTER is not a sideshow in American politics. America has a long and proud tradition of helping both U.S. and foreign citizens with crisis funds. In fact, Haiti itself was one of the first cases where Congress intervened, in the early years of the Republic. Those earliest examples of disaster relief would eventually serve as the legal foundation for the New Deal and the welfare state as we know it today.

In its early days, Congress served as both a legislature and a court, deciding on individual claims for relief. At the start of the Haitian slave rebellion, in 1791, the federal government helped finance the white slave owners in their war against the rebels. Three years later, Congress was funding relief for the white Haitian refugees living in the United States. Once it started helping foreigners, it could hardly refuse to help Americans. Congress later gave funds to people who suffered in the 1812 war, as well as an 1827 fire in Alexandria, Virginia. Soon, earthquakes, floods and even insect plagues were covered.[11] That tradition paved the way for New Dealers to extend government relief to the poor and the unemployed, who were suffering from the disasters of the Dust Bowl and the Great Depression. Conservatives opposed their efforts, arguing that such relief amounted to socialism and pushing in vain for the Red Cross to take care of the poor. The effort to move public opinion behind the new social safety net was determined and creative. John Steinbeck's book *The Grapes of Wrath* was—like Dorothea Lange's photos of mothers feeding their hungry babies—a concerted and moving attempt to win widespread support for federal relief.[12] It wasn't long before the movement that started with soup kitchens led to the creation of Social Security.

Within a generation, disaster relief was so big and so dispersed that it needed to be reorganized. By the 1960s, disaster relief was spread across many federal departments, and that overlap led to the Disaster Relief Act in 1974, and the creation of the Federal Emergency Management Agency five years later. A decade after that, the Stafford Act tried to bring more order and financial discipline to disaster relief. But it also added layers of red tape to an already confused and confusing system.

The political debates around disaster relief are still with us today. How much should the federal government lead, and how much should be left to local officials or private charity? Where do you draw the line between the victims of a natural disaster and those who are suffering from chronic poverty? Should there be limitless funding for disaster relief or does that spending need to be offset by difficult cuts to other budgets? Is all the red tape an excuse for inaction or a necessary safeguard against abuse? In Puerto Rico, these unanswered questions made it much harder to deal with the biggest natural disaster to strike American citizens since the Dust Bowl.

MY FIRST EXPERIENCE OF AN AMERICAN NATURAL DISASTER CAME AFTER Hurricane Sandy in 2012. I arrived in New York the day after the storm passed through, and I spent my time watching and learning. I was officially a partner of the Red Cross, and proudly carried a Red Cross identity card. But as I watched the food relief in the battered Rockaways, on Long Island, I quickly learned that the Red Cross didn't prepare the food. That was the work of a group of very organized church volunteers, who came with huge supplies, cooking trailers and support vehicles. I watched them as they cooked a simple meal of mashed potatoes with chicken tenders and gravy, and found the process fascinating. I couldn't believe I had never heard of the food relief work of the Southern Baptist Convention.

The Southern Baptists are the key to feeding Americans in a crisis: they supply 90 percent of the hot food delivered by the Red Cross and the Salvation Army in any natural disaster in the United States.[13] They do so without directly getting any federal funds, relying on church funding instead. They also avoid paying for the ingredients, which come from the Red Cross. For the last fifty years, starting with Hurricane Beulah in Texas, the Southern Baptists have jumped into disaster zones. That includes international disasters such as the 1973 earthquake in Nicaragua and the 1974 hurricane in Honduras. It was Hurricane Katrina in New Orleans that demonstrated how effective they had become: 21,000 volunteers served 14.6 million meals over 7 months, and purified 21,600 gallons of water. That in turn led to a big spike in the number of volunteers trained for the next disaster.[14]

Disaster operations are central to the Southern Baptists' mission: through disaster relief, they say, they can be "the hands and feet of Jesus to people seeking hope during a time of crisis."[15] The results have been incredibly impressive: a trained volunteer workforce of more than sixty thousand. But the trend lines are not encouraging: that number is down from a high of 90,000 in 2008, and most volunteers are retirees with flexible schedules. It's not clear that the younger generation is interested in or willing to do the same kind of volunteering. So the Southern Baptists have dropped their training requirements, allowing online sign-ups and requiring volunteers only watch short safety videos. That's fine for simple physical work like clearing debris, but not good enough for cooking with big-kitchen equipment.

The Southern Baptists like to call themselves the best-kept secret of disaster relief. And anyone who has seen them work, anyone with experience in American natural disasters, knows they are critical to delivering that relief. Without these elderly missionaries, and their mobile kitchen trailers, there is no food for the Red Cross and others to hand out. They may not be food experts, but their experience of working in the middle of disasters is world-class.

Five years after Sandy, as Hurricane Harvey slammed into Texas, I was still watching closely. Four days after the hurricane made landfall in Texas, I flew into Dallas, as the storm stalled overhead, dumping trillions of gallons of rain. I traveled with two other chefs: Victor Albisu from Del Campo in D.C., and Charisse Dickens from my ThinkFoodGroup. We connected with some local chef friends and drove down to Houston. I stopped at a Target on the way to buy boxes of pasta and jars of sauce to prepare dinner for five hundred people living in a shelter that I'd heard about from the Red Cross. They told me there was a very real need for support. After more than eight hours driving around and around, we never reached the shelter because it was totally stranded by flooding, so we gave up. In fact we almost got stranded ourselves by rising floodwaters. Still, the water would not stop us from delivering the pasta and sauce. We ended up giving away our food to a Houston church. It was right there that I realized we needed to prepare heavy-duty vehicles to deal with these disasters, and there was nothing heavier in my world than a food truck. I had put some into action in Haiti a year earlier, after Hurricane Matthew, and I knew these sturdy old trucks could handle anything.

As soon as we arrived in Houston, I made contact with the executive chef of the George R. Brown Convention Center, Edward de la Garza, who had plenty of experience with disaster relief, including after Hurricane Katrina. At the same time, I found out which company was in charge of the facility: Aramark. I called a local Aramark executive in D.C. and said, "Can you get me in touch with the people of Aramark in Houston? Just to get some information because I'm bringing some cooks with me." I wanted to know if the convention center was a place where I could activate the kitchen. It took Aramark three or four days to get back to me. I was already cooking in the convention center by the time my Aramark friend in D.C. texted me: they couldn't track down their coworkers in Houston.

Feeding spaces such as the convention center may be privately managed but in the case of emergency, that's almost always temporary; the government can take them over. In Houston, there was a woman from the Red Cross who wasn't happy with our cooking and had a lot of power over the food decisions. We had a clear idea of what food relief could be; she had a clear idea of what food relief she wanted to see. So they shut down the convention center kitchen for reasons I still don't fully understand.

We could have served so many people, but the Southern Baptists were coming and perhaps they thought they could easily replace Edward's cooking. My team moved on to a children's hospital, as well as to a restaurant called Reef, which was closed after damage from the hurricane. There we worked closely with chef and owner Bryan Caswell and his wife, Jennifer, and we developed the model we would soon use at José Enrique's restaurant, of making sandwiches in his dining room and hot food in the kitchen. We cooked for three days, making ten thousand meals a day, and the experience was invaluable for our operations in Puerto Rico.

In the meantime, I returned to the convention center to deal with a delivery of chicken I had organized, when a big McDonald's truck arrived. It was a very powerful moment for me. The Red Cross used donated burgers to feed twelve thousand hungry, displaced people.

"How many burgers can you produce an hour?" I asked.

"Five hundred," the McDonald's staff replied.

"That's not bad," I said, impressed with their efficiency and scale.

But feeding twelve thousand refugees resulted in people

having to wait two or three hours in line to get their hands on a burger. McDonald's staff had no idea how to serve so many people. This was not a simple challenge, but the convention center had already solved it, through a kitchen designed to feed huge numbers of people.

Later I watched the Southern Baptists set up a kitchen under the highway right behind the convention center. I was amazed that they were using a mobile kitchen unit, because during Katrina, they had used the kitchen in the convention center. I was also worried about what was being done for the people in the outer areas of Houston. Why weren't the relief agencies activating the Southern Baptists and their mobile kitchens in more remote areas? I also saw them setting up in the parking garage, doing a terrific job, with the Red Cross trucks delivering the food. But we had Twitter and Facebook now, and could communicate easily with people in need. I imagined there and then that food trucks could deliver a hot meal from the Southern Baptists, as long as they knew where the need was.

My head was spinning: if only we could match the food intelligence of a group of chefs with the disaster intelligence of the Southern Baptists.

CHAPTER 3

DISCOVERY

WHY PUERTO RICO? WHAT WAS IT ABOUT THIS CORNER OF THE UNITED States that conspired against its citizens to turn a natural disaster into a man-made one? To answer those questions, you need to understand the living history of these islands. The past is never dead, as Faulkner said, it's not even past; and it's nowhere near past in a place like Puerto Rico.

The original people to live on these islands were the Taínos, whose name means "good" or "noble" in Arawak, and who served as the original model for the "noble savage" caricature. Their legacy lives on in the organization of the island's municipalities, as well as in indigenous names such as Mayagüez, Humacao and Loíza. Their language gave us the word to describe the most destructive force they encountered: the hurricane. Most of the indigenous population died from disease and violence—through slavery and armed conflict—which decimated the population from as many as 100,000 to just 1,500 in the first 4 decades after Columbus arrived on the island, on his second voyage to the Americas. Still, genetic studies suggest that two-thirds of today's Puerto Ricans have Native American roots. Puerto Rican food today reflects this history, especially in the Taíno farming of yucca, sweet potato and corn.[1]

The first Spanish conquistadors shaped the island's future in ways we can still recognize today, more than five hundred years later. The first governor was Ponce de León, who gave his name to the island's second city, and who also colonized and named Florida. It was Ponce de León who forced the Taíno into slavery to work in gold mines, construction and farming. When the local population died off, the Spanish brought African slaves to Puerto Rico to work in the mines and the sugar industry. Over the next three centuries, the Spanish forced as many as 75,000 Africans into slavery on the island. At the same time, Spanish immigrants arrived, first from the south of Spain and later from the Canary Islands and Mallorca, attracted by the sugar and coffee plantations. But for the Spanish crown, Puerto Rico's greatest value was strategic: the governor was a military leader and San Juan was a heavily fortified garrison that defended Spanish territory across the Americas.[2] El Morro castle is now one of the UN's World Heritage Sites and remains the icon of the island, proudly displayed on its license plates.

Puerto Rico was the classic colony: a distant territory, reliant on cheap and slave labor, with strategic military value and no democratic rights. By the late 1800s, it was the second largest exporter of sugar, after Cuba, and it attracted immigrants from other Caribbean islands as well as Mediterranean countries. Slaves remained the foundation of the sugar plantation business, and their descendants still live around the plantation centers, in Loíza near San Juan, and Guayama near Ponce. The rise of coffee drove more Puerto Ricans into the highlands of Utuado and Adjuntas, and it also changed the island's diet, as coffee drove out crops like rice and sweet potatoes. More than a century ago, Puerto Ricans were mostly poor, racially diverse and widely dispersed across the island, as they are today.[3]

The Spanish kept tight control of the island, crushing any uprisings or separatists, as they feared another revolution like that in Haiti. While the Creole elite wanted more freedom and prosperity, they were also proud of their Spanish roots and feared a slave uprising. Spain finally granted Puerto Rico some measure of self-government six months before it ceded the island to the United States, along with Guam and the Philippines, after it lost the Spanish-American War.[4]

The United States wanted Puerto Rico for the same colonial reasons as the Spanish: the island was a great naval base and a cheap producer of sugar. Troops invaded the island in July 1898, and the Treaty of Paris in December of that year left it to Congress to decide its political status. Unlike in Hawaii and Alaska, the U.S. Constitution did not fully apply in Puerto Rico, and the Supreme Court created a whole new category of American-controlled land known as "unincorporated territories."[5] The U.S. wanted to own and govern Puerto Rico as the Spanish had since Columbus.

Why the strange language and legal twists? Because the founding story of the United States is about overthrowing colonial power. How could a nation built on the notion of freedom ever admit that it was now a colonial power? This fundamental lie remains at the heart of Puerto Rico's struggles, and played no small part in the island's suffering after Maria.

As a colony by another name, Puerto Rico had no political voice, and lived under an American governor until 1952. For the first half century of American control, Puerto Ricans had less power than they did in the final months of Spanish rule. Under the Jones Act of 1917, Puerto Ricans became U.S. citizens but they did not have the rights of citizens: they had no vote in Congress or rights to a trial by jury as long as they lived on the island. If they moved to the mainland, they magically earned their constitutional rights. They could, however, be drafted into the military, and they were, within a month of the Jones Act becoming law: many

Puerto Ricans were drafted to guard the Panama Canal in World War One.[6] U.S. rule meant an Americanized control of the island, with the teaching of English, a ban on the Puerto Rican flag and the name of the island officially changed to Porto Rico.

Under American rule, Puerto Rico's coffee exports to Europe collapsed as the crop came under U.S. tariffs. But its sugar industry thrived, thanks to the island's special access to the mainland. Still, that flow went both ways: the island's cigar production was hammered by cigarettes from the mainland. Soon the Great Depression dealt a body blow to the sugar industry, and it never recovered.

After the war, the island attracted industrial investment with its cheap labor and special access to the mainland, giving American businesses their highest profit margins anywhere in the hemisphere. By the early 1970s, Puerto Rico was the biggest producer of clothes for the mainland. But as wages rose, production moved to other low-wage countries, and the NAFTA trade deal in 1994 shifted the balance in Mexico's favor forever. Who needed a colony's access to the mainland when Mexico had easier access and a cheaper workforce? In 2006, the ending of generous tax breaks for Puerto Rican subsidiaries of U.S. businesses triggered a recession that continues to this day, as factories shut down. The economy only worsened after the financial crisis and recession that began in 2008.[7]

Puerto Rico did not win its current version of limited political rights until after World War Two, with the direct election of Luis Muñoz Marín as governor in 1948. At first an advocate for independence, Muñoz Marín later campaigned for autonomy under a so-called commonwealth status. As the nationalists embraced armed revolt and even terrorism, Muñoz Marín shifted his position from political to cultural nationalism.[8] The Cuban revolution helped cement that shift, as Cuban exiles moved to Puerto Rico and the island took on new national security importance in the

Cold War struggle against communism. American tourism moved from Cuba to Puerto Rico, along with rum production.[9] There was little desire to follow Cuba's path to independence.

Today most Puerto Ricans are unhappy with their current political status, and the most popular alternative is statehood, while a minority prefer more autonomy. The most recent poll, just three months before the hurricane, showed a 97 percent majority for statehood, although opponents said the vote was rigged and refused to take part in the poll. Five years earlier, statehood won 61 percent of the vote.[10] Yet Congress will not grant statehood to the island in the foreseeable future because, as with the District of Columbia, Republicans oppose what they see as the creation of two new Senate votes aligned firmly with the Democrats. Instead, some Puerto Ricans say that statehood is most likely to come about by Florida annexing the island. That kind of political upheaval could only take place if the politics of the mainland changes, along with the influx of Puerto Ricans onto the mainland after the hurricane.

Whatever happens, the status quo cannot survive. Well before Maria, the island's financial crisis exposed how unsustainable the colonial system is for Puerto Rico. Its governments have taken on crushing debts that they could not pay through two decades of recession. With islanders paying no federal income tax, the government does not have the same economic power as other parts of the U.S. At the same time, 43 percent of the island lives below the poverty line, and the median income per household is just $19,600.[11] Since 2016, the island's finances—and with them, its government—are in the hands of an oversight board appointed by the president.

That economic hardship pre-Maria was already changing the island's population with each passing year. But the hurricane, and its man-made aftermath, has rapidly reshaped the Puerto Rican people, in perhaps the first of the great population shifts resulting from climate change. Florida officials said that more than 200,000 Puerto Ricans arrived in their state within two months of the hurricane.[12] Those numbers are more than enough to tip the balance in presidential elections: Donald Trump beat Hillary Clinton by little more than half that number in Florida in 2016, and the last governor's race in Florida was decided by an even smaller margin. Other data from FEMA and the U.S. Postal Service show that Puerto Ricans have migrated to all fifty states.[13] That means Maria drove more people out of Puerto Rico than either the so-called Great Migration between 1945 and 1965, or the long economic decline since 2000.[14] In two months, Maria changed Puerto Rico as much as did two decades of economic upheaval.

Historically Puerto Rican migrants to the mainland have not been treated well. The Great Migration led to high concentrations of Puerto Ricans in poor neighborhoods of New York City associated with high crime rates. That in turn led to ugly caricatures in movies and on TV, perpetuating the prejudice that first developed under Spanish rule. According to those racist views, Puerto Ricans were either lazy or ignorant, violent or corrupt.[15] It is much easier to cling to that prejudice than address hundreds of years of colonial exploitation, historic poverty, under-investment and financial instability.

Those attitudes have in turn shaped U.S. policies toward the island in ways that are unthinkable for the rest of the country. Take food stamps. In the early 1980s, after Ronald Reagan vilified "welfare queens," Congress decided that Puerto Rico was costing too much in food stamps. Ignoring the poverty levels on the island, Washington simply capped the level of food assistance. The result is that to receive food assistance in Puerto Rico you need to be much poorer than citizens on the mainland (with around one-third of their net income). If you qualify, you receive around 60 percent less in benefits than people who qualify on the mainland. For a family of three, you need to take home less than $599 a month in order to get food assistance of $315 a month. On the mainland, the same family could take home as much as $1,680 and get benefits of $511. To make matters worse, because the funding is capped, it cannot be expanded in case of a natural disaster like a hurricane.[16] It's hard to imagine a clearer signal from Washington to its colonial subjects: you are second-class citizens.

Food is not a minor way to send a message in Puerto Rico, or some functional way to consume calories: it plays a huge role in Puerto Rican pride and culture, and gives us present-day clues to the island's past. Rice, a central part of many meals, was introduced by the Spanish and cultivated by African slaves in the marshlands.[17] Today's rice or plantain *pasteles,* or patties, wrapped in leaves and boiled in water, are based on African cooking techniques.[18] They feature in Christmas feasts including the Spanish import of the *lechón asado,* or roasted suckling pig, and the African and indigenous mixture of *arroz con gandules,* rice with pigeon peas. Another Christmas dish that defines Puerto Rican food—the *mofongo,* or fried plantain—is named after the Angolan word for plate, presumably because the plantain is first mashed on a plate before frying.[19] Puerto Ricans spend as much as half their income on food, reflecting a combination of the poverty on the island, the high cost of food imports and the cultural importance of food.[20]

If you want to talk to Puerto Ricans, try sharing a meal with them. If you want to tell them you care, try cooking for them.

CHAPTER 4

BIG WATER

IT WAS THE END OF MY FOURTH DAY; A WEEK AND A DAY AFTER THE HUR-ricane tore up the island. We were back at the penthouse bar at the AC Hotel in San Juan. Outside, on a roof with views to the ocean, the pool and cabanas were empty of the usual crowd of tourists and partygoers. Inside, there was an end-of-the-universe atmosphere around a bar that was humming with heavily armed Homeland Security agents, a collection of disaster relief workers, and some locals desperate for power, air-conditioning and a stiff drink.

Even though there had been a dry law in place, I got the hotel to save me a couple of bottles of white wine. I noticed a few Red Cross people sitting on the sofas by the door. Beside them was another group: a Puerto Rican guy sitting with an African-American woman and a buzz-cut man who spoke with a deep Southern accent. He looked like he might be ex-military or ex–Secret Service. They made for a strange group, so I took my wine over to them and started talking about our cooking in Santurce. Three days into our rapidly expanding food relief operation, I already could tell a pretty good story. We had doubled our output from 2,500 to 5,000 meals and we were only just getting started.

"We can feed the island," I told them, as I laid out my plan for operations spanning across Puerto Rico. At the same time, I knew we were already getting squeezed on resources. We were financing all the supplies on our own. My executive director of World Central Kitchen, Brian MacNair, had just arrived on the island and he wasn't exactly relaxed about the size and growth rate of what we were doing, with no money to pay for it all. We had a little cash left over from our last fund-raiser, but it was nothing like enough for what we had begun. We were a small nonprofit with no record of massive fund-raising in a crisis.

The Puerto Rican guy was a lawyer and a politically connected fixer by the name of Andrés López, who was an early supporter of one Barack Obama as president. They were at Harvard Law School around the same time, and López was rewarded for his support by being named a member of the Democratic National Committee and a trustee at the Kennedy Center in Washington. He was sitting next to another DNC member, Karen Peterson, a state senator from New Orleans and chair of the Louisiana Democratic Party. She was recounting the state's early relief and recovery after Hurricane Katrina in 2005.

As I continued to outline my vision for feeding one million people across the island, the Southern guy turned to the other two and pointed to me.

"This is it," he said. "This is the guy. Let's bring him in tomorrow to FEMA."

"What do you mean *this is the guy?*" I asked.

It sounded like they were sitting there waiting for something to happen, or someone to show up. Someone with an idea they could use.

The Southern guy was Josh Gill, and he started to explain his background. He was also from Louisiana, where he served as a state emergency bulk fuel coordinator, and had some experience with the recovery after Hurricane Katrina, as well as some knowledge of the Federal Emergency Management Agency. It wasn't clear if Peterson and Gill were trying to help or looking for business, or some mixture of the two. López was a friend of Peterson, and was staying at the hotel with his family to give them a break from their blacked-out home.

"I'm going to bring you in tomorrow to talk to the FEMA guys," Gill told me. "We'll go to the mass feeding meeting."

We talked in vague generalities about what we hoped would happen: some combination of FEMA and the Red Cross would pay us to feed one million Puerto Ricans. That was only 989,000 meals more than we had already cooked. If it sounded like an impossible dream—both the funding and the feeding—that's because it was.

AS FAR AS THE TRUMP ADMINISTRATION WAS CONCERNED, EVERYTHING in Puerto Rico was going exceptionally well. "FEMA & First Responders are doing a GREAT job in Puerto Rico," Trump tweeted that evening. "Massive food & water delivered. Docks & electric grid dead. Locals trying . . . really hard to help but many have lost their homes. Military is now on site and I will be there Tuesday. Wish press would treat fairly!"

It wasn't clear where he was getting his information about the "massive food & water" deliveries. They certainly weren't obvious to those of us on the ground, or in any of the official reports.

"Puerto Rico is devastated," he continued in another tweet. "Phone system, electric grid many roads, gone. FEMA and First Responders are amazing. Governor said 'great job!'"

At least he recognized how bad the infrastructure was on the island. We'd had many differences between us, but on this I had to agree with him: the island was clearly devastated.

The tweets were probably the result of news coverage of his own staff briefing earlier in the day. Brock Long, FEMA's administrator, gave Trump a personal walk-through of what was going on. It was unclear what was more detached from reality: Long's account of Puerto Rico's situation or Trump's understanding of it.

Either way, it resulted in lots of happy talk from Press Secretary Sarah Sanders at the podium in the White House briefing room. "The full weight of the United States government is engaged to ensure that food, water, healthcare, and other lifesaving resources are making it to the people in need," she began.[1] I live in Washington, D.C. I know what the full weight of the U.S. government looks like. It didn't look anything like the food, water and health care operations in San Juan at the time she said these words. Or at any other time, for that matter. And San Juan was by far the best served place in Puerto Rico.

Once again, the White House proudly declared that there were lots of people on the ground. "Ten thousand federal government relief workers are there, including 7,200 troops are now on the island and working tirelessly to get people what they need," Sanders said. "We have prioritized lifesaving resources to hospitals and can report that 44 of the island's 69 hospitals are now fully operational."

I'm pretty sure that a week after Hurricane Sandy, if only two-thirds of New York's and New Jersey's hospitals were functioning, there would have been street protests, lawsuits and maybe criminal charges. It didn't seem like something to brag about, but rather something to be ashamed of. In any case, there were credible reports that only one-third of the hospitals were functioning.[2]

"There's a long way to go, but we will not rest until everyone is safe and secure," she concluded. "Our message to the incredible people of Puerto Rico is this: The President is behind you. We all are—the entire country. Your unbreakable spirit is an inspiration to us all. We are praying for you, we are working for you, and we will not let you down."

We will not let you down. It was hard to forget those words, so easily repeated, in the weeks to come.

Of course, it wasn't just the communications side of the White

House that was the problem. Accompanying Sanders at the podium was Tom Bossert, Trump's Homeland Security adviser, the top official inside the West Wing overseeing the recovery.

His words amounted to a colossal admission of failure, dressed up as a boast about operations.

"Through aerial surveillance we've seen the entirety of Puerto Rico," he told reporters, admitting that after a week on the ground, the United States had failed to check on its own citizens on an island just seventy-one miles long. "Some of the southwest and southeast sections of the island have had a little bit more sparse on-foot exploration," he said. The southeast was where the hurricane landed.

"But it's the interior of the island that's presenting the biggest problem for us right now," he said. "The mountainous interior is where we're dedicating our efforts to try to get in with rotary wing support."

This wasn't Afghanistan in winter surrounded by Taliban terrorists hiding in caves. The mountainous interior of Puerto Rico was not hostile territory in an impossibly harsh climate. The Trump administration had failed to throw its full weight behind the recovery. After the earthquake in Haiti, a foreign country, eight thousand U.S. troops were en route within two days. A week after a hurricane on U.S. soil, we had fewer troops on the ground, on an island with literally tens of thousands of U.S. military stationed here. The idea that we couldn't mobilize troops to visit the interior of the island until ten days after the disaster was stunning. How exactly did the world's most powerful military force invade foreign countries if it couldn't reach the middle of an island it fully controlled?

And how did the administration explain its vastly different responses to Harvey in Texas and Maria in Puerto Rico? By this time, they had seventy-three helicopters over Houston. It would take three weeks to get that number flying over Puerto Rico. By this time, they had distributed more than twenty thousand tarps in Houston, but just five thousand in Puerto Rico. By this time, they were on the verge of approving permanent disaster work in Texas. It would take another month to do the same in Puerto Rico.[3] And Maria affected so many more people over so much more territory than Harvey.

One reporter standing at the side of the briefing room confronted Bossert with some facts from Puerto Rico.

"Tom, I've got a text here from a volunteer who has boots on the ground and he says that they need helicopters to evacuate people from remote areas of the island," he said. "And he says there are people burying their family members in front yards, communication is badly needed, and they look at apocalyptic conditions between 48 and 72 hours."

Bossert insisted that he wasn't going to "micromanage" the recovery. "That's the mistake you've seen in the past," he said. "I believe—I'm confident anyway—we've got enough resources marshaled and deployed forward to make those decisions under the right command and leadership structure. What we've done, and as I've explained in the past, is we've had to augment and change our business model in the field."

This business model was nothing like any commercial enterprise I had ever seen. If it was a business, it wasn't clear who the customer was, or how the government was serving them. For Bossert, everything was just fine, even for the people texting about their urgent needs.

"First, people seeing 24- and 48-hour horizon problems where they're saying, 'I don't see enough food and water coming,' it's my sincere belief that that food and water is going to get to them before that deadline arises and that we're going to save their lives. I have no doubt in it."

His belief was rooted in the numbers, he said. "We've got commodities distribution now exceeding millions. So 1.3 million meals, 2.7 million liters—that type of thing—of water."

There were 3.4 million people living in Puerto Rico before the hurricane. After eight days, and just two meals a day, they would need 54 million meals. Even if 90 percent of the island had access to its own food in that time, the remaining 10 percent in the greatest need would be 4 million meals short. As for water, doctors say the average person needs around three liters a day, unless he is sitting in air-conditioning all day. That wasn't the case in tropical Puerto Rico, limping along on a few generators. It was shocking that the president's most senior adviser would feel good about distributing 2.7 million liters of water—less than a liter per person for an entire week.

"Now, if there's somebody burying somebody in their front yard, that's an absolutely terrible story," Bossert said. "What I don't want to do though is project it as the norm, and I think there's a careful distinction here."

"What is the norm?" the reporter asked.

"Right now we've seen 16 fatalities confirmed from the state authorities," he replied. "No fatality is acceptable. If that number increases significantly, that will be a devastating blow. We are going to do everything we can to prevent that. The loss of life from the storm is one thing; loss of life that's preventable is another. And that's why we're trying to marshal our resources."

THE JONES ACT OF 1920 WAS ONE OF THOSE PIECES OF LIVING HISTORY that told you everything about Puerto Rico's status as a colony. Its formal name is the Merchant Marine Act of 1920, signed into law just three years after the people of Puerto Rico were granted U.S. citizenship, just in time to be conscripted into the U.S. Army in World War One. The 1920 Act was supposed to protect the U.S. shipping business by requiring that all sea freight travel between U.S. ports on U.S.-built ships, owned by U.S. citizens and operated by U.S. crews. It is hard to imagine a more nationalist and protectionist law, especially at a time of a globalized economy, where sea freight flows through so many international players. But at the time, the Jones Act of 1920 was intended as a way to boost the maritime industry and U.S. national security. It also enshrined the privileged status of U.S. territories like Puerto Rico, Guam, Hawaii and Alaska. The latter two only became full states forty years after the act. In the meantime, these countries all had special access to the U.S. The exchange was clear: in return, like all good colonial powers, the U.S. imposed special control for its favored corporations.

Whatever the original intent and effect, the Jones Act of 1920 is a singular disaster for Puerto Rico today, even without a hurricane or two to make things worse. The Federal Reserve of New York concluded that the Jones Act "does indeed have a negative effect on the Puerto Rican economy," according to its 2012 study of the island's competitiveness.[4] It couldn't say exactly how much of a negative effect that was, but it did make these helpful comparisons. First, it pointed out that just four carriers control all shipping between the mainland and Puerto Rico. Another study revealed those carriers operate just five container ships to supply the island.[5] The numbers are clear in terms of the impact on prices. It costs twice as much to ship a twenty-foot container of household goods from the East Coast to Puerto Rico as it does to nearby Santo Domingo in the Dominican Republic or Kingston, Jamaica. As a protected trade, those prices remained stable in Puerto Rico even when the global glut of shipping was driving costs down elsewhere in the same part of the Caribbean. Puerto Rico was hurting, the New York Fed found. Over the previous decade, the port in Kingston had overtaken San Juan in total container volume, even though Puerto Rico's population was one-third larger and its economy was more than triple the size of Jamaica's. To add insult to injury, the nearby U.S. Virgin Islands had been exempted from the Jones Act since 1922.

If the shipping restrictions were harmful in normal times,

imagine what it was like in the aftermath of Maria's devastation. From the first days after the hurricane, reporters began asking the administration about lifting the Jones Act to help with supplies to the island. After all, the administration had issued a temporary shipping restrictions waiver for Texas and Florida after hurricanes Harvey and Irma just a few weeks earlier.[6] Homeland Security officials told reporters that this was because those hurricanes affected oil supplies, whereas Maria did not affect national security and there were enough American ships to supply Puerto Rico, in any case. This kind of thinking is impossible to understand. Why is gasoline for cars a more important form of energy than food and water for people? As a chef and as an American citizen, I cannot accept that food and water are excluded from national security concerns. Never mind that Puerto Rico is critical for medicine production for the mainland.

Soon journalists were asking Trump himself why he wouldn't waive the Jones Act shipping restrictions for Puerto Rico as he had in Texas and Florida. His answer was more honest than the Homeland Security officials' response: because the shipping industry didn't want it lifted.

"Well, we're thinking about that, but we have a lot of shippers and a lot of people who work in the shipping industry that don't want the Jones Act lifted," he told reporters on the South Lawn of the White House.[7] "And we have a lot of ships out there right now.

"Puerto Rico is a very difficult situation," he explained. "I mean, that place was just destroyed. That's not a question of, gee, let's dry up the water, let's do this or that. I mean, that place was flattened. That is a really tough situation. I feel so bad for those people."

It was nice that he felt bad for those people who were his fellow Americans. But he didn't feel bad enough to overrule the shippers, who were happy with the way things were. Two of the main operators, Crowley Maritime and TOTE Maritime, told the *Wall Street Journal* that their containers were stuck in the port because

of a lack of ground transport.[8] "We just can't move trucks to depots
and gas stations," said one executive who asked not to be named.
"The roads are a mess."

Some of the roads were a mess, but many of them weren't.
When members of Congress demanded the lifting of the Jones
Act, the White House caved—just one day after Trump talked
about the shipping industry's opposition. Defense Secretary Jim
Mattis suddenly found that it was in our national security interests to waive the act. The White House announced the reversal
in its approach through a tweet from Sarah Sanders, just a few
hours before she promised *we will not let you down*.[9] It was good
news, for sure. But it was also a clear sign of how poor the preparations were for this recovery; how little thinking had been done,
beyond protecting some vested interests; and how the response
to Puerto Rico was being cooked up on the spot.

WE WOKE UP EARLY, THE MORNING AFTER MY CHANCE MEETING AT THE
penthouse bar. FEMA's meeting about food and water—coldly called
the "mass care meeting"—was scheduled to start at 7:30 a.m. at
the convention center.

The Puerto Rico Convention Center in San Juan sells itself
as the most technologically advanced in the Caribbean. That may
be true. But technology does not equal intelligence, and information about what was really happening on the island was hard
to find inside its concrete and glass atrium. What you could find
were Americans from the mainland. Lots of them milling around,
traversing escalators, ducking in and out of meetings that had
no visible impact on the life of the American islanders outside.
The only way to know you weren't in any other convention center
across America was the sight of so many armed personnel, police
and National Guard, in camo and flak jackets, bearing the kind
of firepower that could have protected U.S. forces in Baghdad's
green zone. They lived in a perfectly supplied bubble, eating sushi,

drinking beer and playing the slot machines in the Sheraton hotel across the street. Judging by the number of automatic rifles inside, the federal government was far more ready to protect itself against invasion than it was to protect its citizens from a humanitarian disaster. All those guns did not stop a steady stream of volunteer arrivals direct from the airport. You could tell the newcomers on sight: hauling bulging suitcases through the steel doors, wearing brightly branded T-shirts saying things like Love 4 Puerto Rico.

We entered through a back door because we had no official credentials or invitation. Chefs can always find a back door to any building like this: the kitchen staff need to go in and out quickly so there's usually an entrance propped open for a cook to smoke a cigarette or take out the trash. I found that kitchen door quickly and we took the stairs to the floor with all the government meeting rooms.

The mass care meeting took place in a hallway where officials from all the main federal departments gathered—defense, agriculture, housing, education and FEMA, of course—as well as from the big NGOs, like the Salvation Army and the Red Cross. Josh Gill met me there but I already knew Captain Sanderson from the Salvation Army. It was obvious to me that there was no real coordination or leadership. Nobody seemed to be in charge, and everyone was doing their own thing. There was no overview or intelligence about questions concerning the real needs of the people: Where were the people in the worst situations, and how many were they? Who had what resources and how could we combine forces? The meeting was so chaotic, the participants decided to have another food meeting after the mass care meeting, on the balcony outside. "Just so we're clear," said one person, "this is a sub–task force." These were the things they cared about. Sub–task forces.

I just cared about getting more bread, cheese, ham and mayonnaise. I thought our feeding operation was scrappy and still had a long way to go to reach a smooth state. But compared to how we had ramped up at José Enrique's restaurant, the federal government was in complete chaos. There was apparently food and water stored in a government warehouse but the delivery trucks didn't have the right approvals to go there. There was no way to match the words from Washington with what was happening in San Juan. I couldn't see how this group of officials could get anything done. The government was unprepared, more than a week after the hurricane, and that made me feel—more than anything else—very sad. It also made me hate meetings, and endless planning, even more than I normally do. We didn't need a never-ending strategy discussion at José Enrique's. We just began cooking and delivering food.

The session was led by FEMA's head of mass care in Puerto Rico, Waddy González. González was a former Red Cross official who knew the island well, having grown up in San Juan and graduated from the University of Puerto Rico. I had no idea at the time just how close the relationship was between the federal government and the Red Cross; I naively thought the Red Cross was an independent charity, even if it was huge.

But when it comes to what they technically call "mass care," the Red Cross and FEMA are so closely entwined that they co-chair something called the National Mass Care Council.[10] That group produced a strategy document in 2012, and González was a key player in its creation. The strategy calls itself "a road map for national mass care service delivery." It's full of technical language like *standardization of terms* and *expanding capabilities*. It has no section dedicated to food and water. Its recommendations are for more planning and processes, and it makes clear that nobody is really in charge. In fact, it prides itself on that lack of structure. "The strength and resilience of our current system is that we do not rely on a single entity for the provision of mass care services," the strategy document declares, "but have a history of collective action by government, nongovernmental organizations, faith groups, the private sector, and other elements of our society—the Whole Community."[11] This is a great idea in many cases: to be adaptable and open to anyone who can help, especially the private sector. However, the reality was the opposite: rigid, closed, unresponsive and unwelcoming to the private sector mind-set.

González started with what sounded like an impossible goal. He said FEMA had identified that 2 million people needed feeding in Puerto Rico: almost two-thirds of the population. At three meals a day, that meant an operation producing 6 million meals a day, across the island. That number sounded right to me, but those were still eye-popping numbers. Who could produce that much food, never mind deliver it? We had just doubled our output to five thousand meals a day. I dreamed of reaching one million meals in total. It was a colossal amount of food, and the sheer scale of that goal seemed intimidating. But it also felt right: with a lot of help, we could achieve the impossible because the people needed feeding.

My hope was that we could partner with the Red Cross and Salvation Army. I thought we might be able to supply them, to feed into their system of delivery. I was unsure about speaking up, which is an unusual feeling for me. But I was just a chef, running a small and young nonprofit that was never supposed to be there, in a meeting with people representing huge charities. The Red Cross has annual revenues of $2.6 billion. We weren't just small compared to them; we were microscopic.

But Josh Gill prodded me to talk, so I gave it my best shot. Here was my plan: We could source the food and deliver it to kitchens across the island, staffed with chefs and volunteers, making rice and chicken, as well as sandwiches and soup. We could double and double again, every day. Boom, boom, boom. Surely there was some way we could partner to feed the island? I looked around. The group seemed to like my energy but that was about it: they looked at me like I was a smart-ass with some crazy vision of saving the world.

"You and I need to work together," said Sanderson from the Salvation Army. "I'm going to take care of you."

I thought, *Great! He's going to give us the money. I don't need to worry.* In hindsight, it was clear he actually meant that I would just give him the food. They avoided any further talk about money.

"Who is going to distribute the food?" asked González.

"Everybody who has a car," I said, knowing that municipal officials were already coming to pick up the food in Santurce.

"What if they don't deliver?"

"Let's say 10 to 20 percent of the food is stolen," I replied. "So what? They will still eat the food!"

The Red Cross seemed annoyed by my plan and asked the kinds of questions that made me feel they didn't really believe we could stand up to anything. I turned the question back on them.

"Where are the trucks of the Red Cross?" I asked. "When are the units of the Southern Baptists coming? Where are the red Cambros for distributing the food?"

"We have no trucks," they said. "And we have no units of the Southern Baptists."

"If you don't have trucks and the Southern Baptists, you can't feed the people," I replied.

You also couldn't feed the people if you didn't understand the island. It sounded to me like they had no idea where the food was produced and stored on the island. For instance, I had been talking to José Luis Labeaga, the owner of the Mi Pan bakery, who told me they would donate bread and contact the other bakeries to do the same. They had enough flour and frozen bread to keep donating bread, and it was produced by local people. The only catch was that we needed to send people to pick it up because they had problems with their trucks. Did FEMA and the Red Cross know there were working bakeries in Puerto Rico, which were desperate to help? If they supplied diesel and gas to the bakeries, they could have helped to feed the island.

"Why don't you get a gas tank to every bakery so the bread is delivered around the island?" I suggested.

"We have other priorities," said González. "Food isn't a priority right now."

I couldn't believe what I was hearing, and they couldn't believe what I was saying. If I were president of the United States or the director of FEMA, I would make food and water the top priority. With a couple of pieces of bread, you can easily put something in between and make a good sandwich. In a moment of real need, a simple sandwich looks like heaven. And if you feed the people, you are creating an army of first responders. If you look after people in their time of need, they become the most important and effective response: they become volunteers.

González looked at me carefully, like he was trying to comprehend which planet I came from or what language I was speaking. He seemed to be struggling internally with wanting to help while also wanting to say no.

"José, you don't understand the process," he kept telling me. "We can't do this as quickly as you want."

It wasn't about me wanting to do things quickly. I just wanted to feed people as quickly as they needed food, which is to say, every day. It felt like González had been at FEMA for so long that there was only one way he could look at the world. He was a sympathetic man, sensitive to the needs of Puerto Rico, but he seemed handcuffed by the system. I had heard that he was concerned about his family on the island, but was too busy to go see them. He clearly placed his sense of duty and public service above his personal needs. But he also could see no way to cut through all the red tape to help the Puerto Rican people he knew so well.

"We'll see what we can do," he said.

As for the Salvation Army, their situation seemed hopeless.

They had a kitchen running in Ponce, on the southern coast, but were struggling to produce any real quantities of meals. They complained they were having trouble getting cash out from the banks so they were flying cash into the country. As a result, they were finding it hard to buy enough food to cook.

They complained they could only bring in less than $10,000 at a time, which confused us. This wasn't a foreign country and there were no restrictions on cash, because there were no customs. It was still the United States.

By this point, we had already spent more than $80,000 on food through our suppliers, José Santiago, based on a line of credit that was opened with a simple handshake. The Salvation Army, as huge an organization as you could imagine in this sector, was struggling to spend $10,000. They seemed paralyzed by the whole mission, and could barely cook two hundred meals out of their Ponce kitchen. That was a full digit less than we were cooking out of a leaking restaurant and car park after a couple of days of effort. They asked for our help getting their kitchen up to speed, but it sounded like more bureaucracy than it was worth, and their kitchen didn't have the capacity we clearly needed to feed the island.

"What the hell is going on here?" I said to my friend Nate. "The Salvation Army can't buy food?"

On my way out of the convention center, I noticed a military team whose job it was to create maps. My dream was to get an overview of the island—a comprehensive summary of every area of need, showing every bridge that was down, every gas station that was open or closed, and every community that needed most help. It was only with that kind of information that we would be able to locate our kitchens, or target our deliveries, in the right parts of the island at the right time. I started talking to the mapping guys and one of them said he'd been to my restaurants in D.C. That day they were running out of ink for their maps, and

I promised them I would try to get some new supplies for them. They were modest people—not what you would expect from the self-confident American military—and they seemed to appreciate my offer of help.

"I'm going to be feeding 100,000 people a day—maybe more," I told them, to help them imagine the scale and detail of maps that I'd need.

"Wow," was all they could say.

I knew I sounded like a crazy dreamer, but that was a risk I was willing to take. Sometimes when I speak my mind, my dreams help push me—and those around me—to achieve more than we thought possible. It's never good to over-promise and under-deliver. But in this crisis, to produce any meals would be better than the current catastrophe.

After the meeting, I regrouped with Josh Gill and we wrote to the Red Cross detailing what food supplies we needed to fulfill a big order for 240,000 sandwiches, including five thousand pounds of deli-sliced ham and bologna. "Let me know when we can get them," I wrote. Based on their disinterest, we weren't holding our breath. We needed to give them the chance to step up, but it didn't sound like they knew how to manage a food operation in these conditions. We were happy that World Central Kitchen, our small nonprofit, was at the table for the talks with the biggest NGOs like the Red Cross. And who knew? If they couldn't get their hands on the supplies, they at least had the resources to pay for the sandwiches.

The conversation with Gill was an eye-opening insight into the world of disaster relief: a jungle of tenders and requirements, contracts and fees, of middlemen and bureaucrats handling millions of dollars. It was a world away from the hungry, thirsty Puerto Ricans just a few minutes' walk outside our door.

Gill said he could help get us a food contract from FEMA and I was naïve. I thought everyone was working out of the goodness of their heart, but it was clear that Gill expected more, so I asked him what he was going to get out of it. He named his price: one dollar per meal would go to him. I had no idea who he was barely twenty-four hours earlier, but he had already proved his value to me. He took us to the right people in the right meeting, and they were comfortable with him. Above all, he gave me the courage to speak because he believed in me. Even if he believed in me because he hoped there was money to be made.

My head was spinning with all the jargon and numbers. I just wanted to feed the people. At first, I said okay to his price: it seemed like some combination of Monopoly money and The Hunger Games. It was only later that I realized how much money this could stack up to be, after a few hours of work by Gill. "This is fucked up," I told Nate. We briefly considered halving his fee, to 50 cents a meal, thinking it was a tough piece of negotiation. But then we caught ourselves: we were getting sucked into a crazy calculation that would take food away from hungry people. If we cooked one million meals, as we hoped, Gill would still make out with half a million dollars.

I called him back to put a cap on his earnings at $250,000. Gill was upset with me, but I had to insist. This was not the time or place for anyone to get rich. I wasn't interested in contracts and didn't understand this business. I only wanted to feed the island.

WE DIDN'T HAVE MUCH MONEY, BUT WE DID HAVE SOMETHING PEOPLE wanted: food. We were improvising and doing whatever it took to get that food delivered. So we weren't above bartering the food for an essential commodity: gasoline. Ricardo Rivera Badía of El Churry had a brilliant idea: one of his restaurants was right next to a gas station, and he knew the owners loved his food. So we traded food, and lots of it, for as much gasoline and diesel as we needed. It was one of our very best deals, and it allowed us to feed so many more people across the island. When our mobile kitchen ran out of gas, we drove to this gas station to fill ten plastic cans of fuel, some of them with no caps to stop the gasoline from sloshing out. We literally traded food so we could carry on cooking some more. FEMA asked us how we were going to distribute the food. The answer was simple: with creativity. If people know you are cooking, the right people will always show up to eat and help deliver the food to others.

That day our cooking operation was bigger than ever. We were still cooking huge amounts of *sancocho,* as well as making sandwiches inside the main dining room. Outside, in the car park, we now had three giant paella pans cooking chicken and rice. On top of all that, we had three food trucks reaching out farther into the communities of need, both inside and beyond San Juan.

In our mobile kitchen, parked in the street opposite José Enrique's pink restaurant, chef Wilo Benet made *pastelónes* for the first time, a lasagna-like specialty of the island, complete with layers of sweet plantain, beef, potatoes, and tomatoes. Wilo was one of the great chefs of Puerto Rico and his Pikayo restaurant was a trailblazer. But it was now shut, along with the Hilton hotel that was its home. Rather than waiting for insurance claims to be processed, he was one of the first to join us at José Enrique's. He cooked in unbearable heat, but he didn't seem to mind. Like the rest of us, he knew that he was sweating to help the whole team serve the greatest needs in Puerto Rico. "It's a good feeling knowing you are doing something to help," Wilo used to say.

Our second kitchen was humming at Enrique Piñeiro's Mesa 364. Taped to the wall inside the restaurant was a flip-chart sheet that read just like the ones inside José Enrique's restaurant. The places and numbers told the story of an island in desperate need: 175 hot meals for the elderly at one *egida,* 75 at another and 100 sandwiches for a third. A tray of meals for fifty at the children's hospital, another for the medical students at the University of Puerto Rico hospital, yet more for a nursing home. And barrio after barrio asking for hundreds and hundreds of hot meals. We weren't in the business of saying no. We took every call that came in.

If anything, my problem was that we were growing too fast. Enrique wanted mayonnaise and cheese for more sandwiches but I insisted that we centralize our sandwich-making. The only way we could scale up this kind of thing was to focus on big production lines. "We want to use 100 percent of the donations we get," he texted me. "We have pastrami, legs of ham and bread. What do you propose we do?"

"Bring them to Santurce!" I shot back. "Donations for donations don't work. We have to centralize. We will create a central headquarters for sandwiches. Use your bread to accompany your rice and chicken!"

The contrast with the skepticism, confusion and inertia at FEMA headquarters could not have been greater. We hit 10,000 meals that day, including 2,000 sandwiches, to bring us to a grand total of 21,500 meals cooked in four days of operations. We had doubled our output and doubled it again. There was no reason to think we would stop here.

That night my friend Anderson Cooper was broadcasting his CNN show live from outside a gelato shop at the Marriott, just a few blocks from our hotel. I told him I cried a lot for Puerto Ricans that day; I cried for their suffering, for their generosity and because I felt we all needed to do more. I believed we all could do so much more. Anderson talked about seeing families cooking outdoors for entire neighborhoods, and it was true. The spirit of Puerto Rico was always astonishing. But it was especially powerful at a moment when others thought the worst of Puerto Rico, or would have behaved worse themselves in such a disaster zone. It was time for the United States to step up like its own citizens in Puerto Rico.

"What you see is the heart of Puerto Rico," I told Anderson. "When there are moments of hardship, they come together. And with nothing, they are able to do a lot."

Anderson said it was sad that so many Puerto Ricans felt the need to say they were American.

"This is very simple," I said. "I've been in Haiti for many years. I went after the earthquake. And the amount of help that came from America was far and away bigger than the amount of help that has come to Puerto Rico from the military. At one point we had 25,000 military in Haiti and we're not very close to that in Puerto Rico. So this message is very simple: Mr. Trump, we want you to lead. But let's keep on doing what we did in the past so successfully. Not different, just equal."

It wasn't clear what Trump was doing about Puerto Rico, judging from his public comments. Earlier in the day, he told a meeting of the National Association of Manufacturers about what he called "a massive federal mobilization" of five thousand military personnel.

"All appropriate departments of our government, from homeland security to defense, are engaged fully in the disaster and the response and recovery effort—probably has never been seen for something like this," he said, forgetting entirely about Haiti.[12]

"This is an island surrounded by water—big water, ocean water."

Big water, ocean water. Puerto Ricans would repeat these words for weeks to come, giggling at the president's understanding of their plight.

Worse, Trump thought the island and its government were obliterated in the apocalypse, yet he was still boasting that five thousand military was sufficient.

"We're closely coordinated with the territorial and local governments, which are totally and unfortunately unable to handle this catastrophic crisis on their own. Just totally unable to. The police and truck drivers are very substantially gone. They're taking care of their families and largely unable to get involved, largely unable to help. Therefore, we're forced to bring in truck drivers, security, and many, many other personnel, by the thousands. And we're bringing them onto the island as we speak. We've never seen a situation like this.

"The electrical grid and other infrastructure were already in very, very poor shape. They were at their life's end prior to the hurricanes. And now virtually everything has been wiped out, and we will have to really start all over again. We're literally starting from scratch."

They weren't literally starting from scratch. Everything wasn't wiped out. But it would have been great if the Trump administration behaved like it were. Instead, what we saw was a long way short of the world's most powerful nation responding to "this catastrophic crisis."

"We will not rest, however, until the people of Puerto Rico are safe," he promised. "These are great people. We want them to be safe, and sound, and secure, and we will be there every day until that happens."

I hope that the United States will be there every day. Puerto Rico is, after all, part of the United States.

At the same time as Trump was describing the big water around Puerto Rico, the mayor of San Juan was describing the real situation on the ground. While Trump was claiming to lead, Carmen Yulín Cruz was demanding some actual leadership from the federal government. Her urgency was not just the result of what we were all seeing on the island; it was a response to the White House boasts, the day before, about the size of its operations. *We will not let you down.*

"I am asking the president of the United States to make sure somebody is in charge that is up to the task of saving lives," she told reporters. "They were up to the task in Africa when Ebola came over. They were up to the task in Haiti. As they should be. Because when it comes to saving lives, we are all part of one community of shared values. I will do what I never thought I was going to do: I am begging. I am begging anyone that can hear us to save us from dying. If anybody out there is listening to us, we are dying. And you are killing us with the inefficiency and bureaucracy."

Her words about Ebola rang true. My friend Ron Klain was named the Ebola czar by the Obama administration and his work was an example of how best to handle an emergency that nobody had planned for. Ron's leadership was vital in the successful turnaround as the world helped to control and eradicate the Ebola outbreak in West Africa in 2014.

Pointing to supplies behind her, the mayor said: "This is what we got last night: four pallets of water, three pallets of meals and 12 pallets of infant food. Which I gave to Comerio, where people are drinking out of a creek. So I am done being polite. I am done being politically correct. I am mad as hell because my people's lives are at stake. And we are but one nation. We may be small, but we are huge in dignity and zealous for life.

"So I'm asking members of the press to send a mayday call all over the world. We are dying here. And if we don't stop and if we don't get the food and the water into people's hands, what we are going to see is something close to a genocide. So Mr. Trump, I am begging you to take charge and save lives. After all, that is one of the founding principles of the United States of North America. If not, the world will see how we are treated. Not as second-class citizens, but as animals that can be disposed of. Enough is enough."[13]

Her tone was much harsher than I would ever be, but I understood the passion and the despair. There wasn't going to be a genocide, but she was right: people were dying because of inefficiency and bureaucracy. Puerto Ricans were not being treated as Americans, but as second-class citizens whose lives mattered less than those of their fellow Americans on the mainland.

In fact, the military leadership in charge of the recovery shared her view that the operation fell far short of what was needed. Lieutenant General Jeffrey Buchanan, the three-star general who was named that week to lead the military support for FEMA, said there weren't enough people and assets to deal with the humanitarian crisis. Buchanan, the commander of the U.S. Army North, told reporters bluntly, "No, it's not enough, and that's why we are bringing a lot more."[14]

After my CNN interview I officially met Mayor Cruz for the first time, as well as the Puerto Rico governor, Ricardo Rosselló. They also were being interviewed by Anderson, and both seemed familiar with what we were doing. Cruz was very emotional as she told me of the desperate need for food and water. I promised her we would give her the meals she wanted if she came back to Santurce to pick them up the next day. She started crying and we hugged. She never came over to pick up the meals. So the next day, we got a pickup truck and took one thousand meals to her. For some reason, we never heard from her, or her team, again. We were feeding many people in the San Juan area, which was her area of control. Mayors are important leaders across the island's municipalities, and the success or failure of a feeding operation depends a lot on their leadership. I assumed she was doing fine without coordinating with us, but I was confused by the lack of contact.

Before sunrise the next day, Trump made it clear he wasn't serious about coordinating with the local governments. In a series of tweets directed to Mayor Cruz, he decided she was his political enemy, along with the rest of Puerto Rico, which he cast as lazy and incompetent. "The Mayor of San Juan, who was very complimentary only a few days ago, has now been told by the Democrats that you must be nasty to Trump," he wrote. "Such poor leadership ability by the Mayor of San Juan, and others in Puerto Rico, who are not able to get their workers to help. They want everything to be done for them when it should be a community effort. 10,000 Federal workers now on Island doing a fantastic job."

They didn't want everything done for them. Cruz insisted she was just asking for help, like so many others.[15] It was only "nasty to Trump" because he thought any criticism of him was nasty. What was truly nasty was life in Puerto Rico, where the challenges were so much bigger than one man's ego, even if that man was the president of the United States.

Trump's comments earned a swift and searing response from Lin-Manuel Miranda, the musical genius who created *Hamilton* and *In the Heights,* and whose family is from Puerto Rico. "You're going straight to hell," he tweeted at Trump. "No long lines for you. Someone will say, 'Right this way, sir.' They'll clear a path."

BY SATURDAY, ONLY FIVE DAYS AFTER I LANDED IN PUERTO RICO, IT WAS clear to me that we had outgrown our space in Santurce. We needed to go back to our initial plan: the giant facility of an arena, with its huge kitchen, staging areas, delivery bays and road access. We needed to get inside El Coliseo, known to everyone as El Choli. That space was in the hands of the first lady, the governor's wife, who was using the main arena floor to store donations of food, household goods, generators and even toys to distribute around the island. It looked mostly empty, but there were still plenty of supplies that could help many people. I didn't understand what she was doing with all those supplies, but I also didn't want to criticize her publicly because I needed her help. The surrounding space that I wanted was totally unused. So among my many conversations, by phone and text, I called the governor's representative in my hometown, Washington, D.C. Carlos Mercader, the executive director of the D.C. office, was a miracle worker. I told him of my hope of getting FEMA funding for our food relief, and our need to go into the much bigger space of El Choli. We also got early support from the wife of the island's secretary of state,

Margarita Rivera, as well as support from Clara Roman, a Puerto Rican entrepreneur. Luvian Rodriguez, a friend of the governor's wife, came along with them. For me, their visit to our Santurce operation was an important sign of moral support, and gave us more connections to the government, and more good references for our work. Within hours, both the arena and the FEMA funding were in motion.

Mercader's ability to break through was a revelation to me. It underscored how nobody was in charge, and everybody was in charge. The governor's office had the power to get things moving, but no apparent capabilities to feed the people of Puerto Rico. The resources were there but you needed to know how to find them and use them. It was as if somebody invited you for dinner and told you to order anything you wanted, but didn't give you a menu. Can I get water? Yes! Can I get rice and beans? Yes! But you don't know you can get sushi and cassoulet. You need a master's degree in FEMA to understand what resources they have and how to make them work for the people who need them.

It was a rainy day but that didn't stop the long lines of people waiting patiently for *sancocho* under their umbrellas in the historic streets of Santurce. And it didn't stop the weekend crowds at each stop made by our three food trucks, which all delivered one thousand meals on their own. We prepared 13,000 meals that day, including 10,000 hot meals: 5 times what we cooked at the start of the week. We were unstoppable.

We now had enough public support that the operation threatened to spiral out of control. With the best of intentions, it's possible to overwhelm a fast-growing operation like ours. People want to help, but they can quickly undermine your best efforts. The visitors at our parking lot were out of control, and I was getting nervous: I even kicked out José Enrique's father, Pepe, because I didn't recognize him. He was proud of what his son had helped to create, but I was worried about all the visitors. I felt so bad. After that, we printed stickers to identify our volunteers and approved visitors.

Our visitors were growing in stature, as well as numbers. The regional head of FEMA, Alejandro De La Campa, came to see what we were doing, and ate a sandwich. He seemed impressed with both the cooking and the food. "Help us to help you feed the island," I said. He remained non-committal. The same day, Governor Rosselló was supposed to show up. We waited for hours but he never made it. I looked out at our incredible start-up kitchen and was almost heartbroken that we had to leave this magical corner of San Juan. Here I could manage and control with my own eyes and hands the whole operation. Soon we would embark on something much bigger that I couldn't see and touch every day. Our time here had been intense. It was only a week, but every hour felt like a day, and every day felt like a week. We were doing so much, for so many people, in such a short time. The sheer volumes were overwhelming.

Ramón Leal from ASORE told me he had received some huge donations: fifteen thousand pounds of chicken, propane gas tanks, hot dog rolls, tortilla wraps, fresh fruit and five pallets of water.

"Ramón, we need to coordinate now!" I told him. "It's too much to move all of that!"

"I know," he replied. "That's why you need to come with us to the radio station this morning. I'm calling out to all my friends, the restaurants and food suppliers, to join in and help." Radio was the most powerful form of communication on an island with very limited cell phone service, but we still needed to take our time and prepare properly.

"Ramón, be careful! I need organization and this can become a circus. Please don't do it yet. Please."

We met at my hotel and talked through a solution. Ramón really wanted to help his relatives in the mountains, so he organized a delivery there. But he was dedicated to helping all the people of Puerto Rico and it was taking its toll on him, as it was on all of us. Like us, he found this food relief operation both exhausting and inspiring at the same time. He was working with other business leaders in the private sector and government officials too. He was bringing a plane full of supplies donated by the Mexican government from Miami to San Juan, with food for our operation and medical supplies, like oxygen tanks. At night, he would deliver diesel and propane to homes for the elderly and hospitals. He would show up and disappear like a ghost, but was always helping somebody, somewhere. He was sleeping just three or four hours a night and had lost over fourteen pounds in the ten days since the hurricane.

"But I never felt better," he said.

WEEKENDS ARE USUALLY A DAY FOR FAMILIES AND FOOD IN PUERTO RICO. People travel to the hills to eat huge meals of roast pork, plantains and rice, at a *lechonera,* or barbecue, where roast pigs turn on vast spits above open fires. They line up at *kioskos* near the beach, where they deep-fry vast numbers of *bacalaitos* with salt cod, or plantain *tostones.* This is the time for family and friends to come together as a community, in a town square, in front of a church, or just at the biggest house in the neighborhood.

Two Sundays after the hurricane, we wanted to support that tradition. We wanted to bring food back to the heart of these hurting communities. Not just to feed people, but to tell them that the outside world cared for them.

It was also our last day at José Enrique's restaurant in Santurce, a place I loved for all its warmth and community spirit. It felt like my home and hearth, where we sweated and struggled our way to cook tens of thousands of meals. In these narrow streets where San Juan used to party, we had built the foundations for an island-wide food relief operation, and a model for future disasters. But we had grown too big for the space, and as the team prepared to move on, we took our meals onto the roads.

We now had four giant paella pans cooking vast amounts of rice, chicken and vegetables, along with our famous *sancocho* and a huge sandwich operation. We would hit a new record on this Sunday: 20,000 meals, almost 10 times our starting point less than a week earlier, and twice what we had produced just two days ago.

Our first delivery was one thousand meals to Cataño, to the east of San Juan. Under a giant white canopy, with music blaring, we set up tables and served our meals in a party atmosphere. Alongside me were the mayor of Cataño, Félix Delgado, and the island's secretary of state, Luis Rivera-Marín, who were only too happy to spoon out the rice and chicken. You could see the commitment in Félix's face: like many mayors, he was very hands on. Our only problem was that we forgot to bring serving spoons with us. But when you're committed to helping people, you always find a solution, and we survived without them. Four big spoons showed up from I don't know where.

As we set off for our next delivery, our team back in Santurce was packing up the whole operation and moving to our new home at the Coliseum arena.

It was clear from our travels that we needed to grow, and we needed to do that quickly. On our drive to Ponce, in the far south of the island, the devastation was clear. The trees were stripped bare of their branches and leaves. The sad journey felt like we had entered another world, where the trees had just lost an epic battle against some immensely powerful force. Cell phone service died on the way to Ponce and we lost all contact with our team back in San Juan. I had to remind myself this was the United States, not some third-world country, eleven days after a hurricane.

Slowly, some essential services were coming back. Thirty-six percent of Puerto Ricans now had cell phone coverage, according to the governor. One thousand more troops were arriving, but the total number now stood at only 6,400. At the same time, there were 8,800 American citizens in refugee shelters across the island.

Even with expanded numbers of military personnel, the troops were overwhelmed by the sheer scale of the challenge.[16] Besides, those numbers weren't nearly as important as the 55 percent of Puerto Ricans who didn't have access to clean water. That represented 1.87 million Americans without the essential ingredient for life. Fully 95 percent of islanders still had no access to power.[17]

As soon as we arrived in Ponce, we met the city's mayor, Maria Melendez Altieri, known to all as Mayita. She was happy to see us, even as she was troubleshooting an endless list of challenges with minimal support from the San Juan government. When the island's secretary of state showed up at the same time, she calmly took him to task. FEMA had given her a satellite phone that did not work. How was she supposed to handle all the problems of the island's second biggest city?

I already knew about the gravity of the food crisis in Ponce thanks to my friend Lymari Nadal, the actress and writer, who had visited earlier to check on her family in the city. She brought 500 meals but said she could have handed out 5,000 or more.

Together with the mayor, we visited a school that was also serving as a *refugio,* or shelter. I was amazed at how good the conditions were in the kitchen and cafeteria. They had enough food and refrigeration to feed not just the *refugio* but the *egidas* nearby. I gave an impromptu speech, thanking them for the job they were doing, and telling them how we were already cooking thousands of meals a day. They cheered with joy: at times like these, any message of hope is an important boost in the face of such overwhelming challenges.

It was getting late and we moved on with the mayor to some outer areas, where Mayita told us they still didn't have the full picture of how bad the conditions were. Her team was simply overwhelmed by the disaster. That was what we found in the El Tuque area near Ponce, in the southwest of the island. The people there had no water and would walk thirty minutes each way to get their hands on a liter. Some were drinking from a nearby stream out of sheer desperation. There was no power, and the mosquitoes were swarming. The local supermarkets were empty. Our arrival seemed to lift their hopes, even if they initially thought we were FEMA officials coming to save them, because we arrived with an escort of HSI officers. As we served our chicken and rice, along with half an avocado for each person, those we fed were smiling but their patience and good nature masked real hunger and need. We served one thousand meals and took one thousand sandwiches to another community nearby, and we made a critical decision right there: we would open up a satellite kitchen in the area to produce ten thousand meals a day to help serve this community of fifty thousand people, for at least a week or two, until they got back on their feet. Mayita thanked us but I was making promises when I didn't know how I could keep them. Still, I was determined to make it happen.

THAT DAY I WAS STILL FIRED UP ABOUT DONALD TRUMP'S ATTACKS ON Mayor Cruz. "If I was @realDonaldTrump I would be in Puerto Rico to lead no more than 2 days after the disaster," I tweeted. "If I was @realDonaldTrump I would not attack a leader that has worked non-stop for her people," I wrote in another tweet, posting a photo of Cruz. In a third tweet, I said I would praise the volunteers, and in another I said I would stop attacking the media, if I were him. Finally, I suggested he should activate all the food trucks on the island, to create block-by-block kitchens and food delivery. You didn't need a federal bureaucracy to figure it out; you just needed to see what we were doing. Yes, it could look like I was trolling Trump. But my message was deadly serious: we needed real leadership from the White House, not a series of mean-spirited posts on social media. I thought about calling Ivanka Trump: I knew her and her brothers from several encounters, including the litigation around the hotel restaurant that I refused to open with them after their father's comments about Mexican immigrants. I had great relationships with Republicans in the Bush White House and Democrats in the Obama White House. But this administration was different, and I felt disconnected from them.

My argument wasn't with all federal officials. I was clear that I wanted to make things work with the military, with Homeland Security and with FEMA. These were the people who could get things done, and I still had high hopes for what FEMA in particular could do. I thought Brock Long, the FEMA administrator, struck the right note when he appeared on ABC's *This Week* that Sunday morning. When George Stephanopoulos asked him about Mayor Cruz's criticism, and Trump's suggestion that Puerto Ricans were sitting back, he took a long time to clear his throat. "So the success of a disaster response is predicated on unity of command," he said. "The bottom line is, we had a press conference from the joint field office in San Juan. That operation has hundreds of people in it working around the clock to set the strategic objectives. FEMA, DoD, the governor's objectives. We have been working with mayors all around Puerto Rico to make sure we have a strategy." Brock pointed out that Cruz had only been to the joint field office once.

There are very few heroes in any disaster, and there are no perfect leaders. Cruz had flaws, and so did FEMA. Some of those flaws were obvious at the time: Cruz was better at appealing for help than managing logistics. Some of those flaws, especially at FEMA, would only fully emerge much later. For all their bureaucratic checks and balances, the agency was sloppy about its contracts and detached from reality. Still, they had the power to do a huge amount of good for people in desperate need, and I wanted to encourage them to do just that.

Long told ABC News that they were making "slow progress" in Puerto Rico and that the island still had "a long way to go."[18]

"I believe FEMA will make it happen," I tweeted back at Long, after his interview.

None of those details would stop Trump from tweeting about what he considered to be a great success, at the same time as smearing anyone who dared to tell the truth about what was happening on the island. "We have done a great job with the almost impossible situation in Puerto Rico," he wrote. "Outside of the Fake News or politically motivated ingrates, people are now starting to recognize the amazing work that has been done by FEMA and our great Military."

I had no problem recognizing the great work of some people at FEMA, as well as the military. But I could also see very clearly that the administration was not doing a great job. The situation, as we found it, was not "almost impossible" unless you were stuck inside a government bunker with no contacts, no expertise, no local knowledge, and no urgency or creativity about how to feed the people.

ON OUR WAY BACK FROM PONCE, I FELT THE URGENCY OF NOW. THE SITU-ation was so bad in El Tuque that we needed to mobilize the entire private sector quickly. I called Ramón Leal to organize a meeting of business leaders at our hotel. With or without a FEMA contract, we couldn't stop. I was barely eating myself, but I couldn't stop thinking of the people who were drinking rainwater and going to bed hungry that night. The turnout at the hotel was impressive: even the Red Cross showed up. I told them about our plan to feed as many people as possible, and asked them to prioritize donations of food. If they could afford a cash donation, that was awesome too. But I needed to run the operation like a professional restaurant kitchen, and I needed their support. Among the business leaders was Rafael O'Ferrall, a retired U.S. Army brigadier-general, who was now general manager of Dade Paper Company in Puerto Rico. One of our biggest challenges was how to deliver food. Since we weren't sure if we could get our hands on reusable plastic Cambros, we needed a huge supply of aluminum trays.

He took our plan seriously and we would never run out of trays for the weeks and months ahead, as we sent tens of thousands of trays packed with hot food across the island.

After the meeting ended, I wanted to go to José Enrique's restaurant to recharge. But my team was all gone, as they worked tirelessly to move our operations from the small pink restaurant to the giant concrete arena. They were exhausted but they never stopped working. We all were driven by the need to feed an island whose suffering we were still only beginning to understand. I went to the hotel rooftop and lit a cigar, accompanied only by the buzzing of the city's air-conditioning units and generators. As I looked up to the stars, I began to cry. I thought the one star that was missing was the Puerto Rican star on the American flag.

CHAPTER 5

IN THE ARENA

IT WAS ONE WEEK SINCE I ARRIVED IN PUERTO RICO AND THE FIRST DAY in our new home at the Coliseum. We knew we were going to take a big dip in our cooking output as we moved to this huge new space, having hit a record in the number of meals we prepared the day before. It takes time to build an operation like ours, and the task of rebuilding is no easy job.

It was even harder when the largest kitchen on the island was out of action. My experience in Houston underscored what I already understood: that arena kitchens are perfect for a disaster. They are sized to feed tens of thousands of people, have great storage and refrigeration, and enjoy the best access for deliveries going out and coming in. But as I tried to activate the central El Choli kitchen, I encountered challenge after challenge. In my mind, I was already dreaming that we could feed an entire island: we had a kitchen that could feed twenty to thirty thousand people a day. When we arrived there was no electricity and the arena staff seemed like they thought they were doing us a favor by allowing us to be there. *What a waste of space,* I thought. Even the convention center, where FEMA was headquartered, could produce ten times more food from its giant kitchen.

I was determined to get this place up to full capacity, but it wouldn't happen on day one. Still, those initial problems could not hold me back. With the space and facilities we now had, I could see my dreams coming true. I wanted to get to 100,000 meals a day by the end of the week. I liked to think of El Choli as the biggest restaurant launch in the world. We were firing up a huge capacity kitchen that would feed, every day in the open air, 100,000 customers. Perhaps I had that vision because I was also launching a new restaurant back on the mainland at the same time. My team kept sending me photos of the dishes they were creating at our new restaurant Somni at The Bazaar in the SLS Hotel in Beverly Hills. One minute I was looking at photos of delicate avant-garde creations for luxury diners in Los Angeles; the next minute I was looking at a giant paella pan of chicken and rice for hungry Puerto Ricans. It was not easy maintaining two lives. When I called my wife, Patricia, to let her know I would not be home any time soon, she naturally asked how long I would stay on the island. Her simple question set free all the tears I had been holding back. I couldn't speak, and she understood.

Our food trucks showed up early at the car park outside El Choli, but I was worried about how long it would take to ramp up the sandwich line. It made much more sense to keep the sandwich operation humming at José Enrique for one more day. We were short of bread as we switched over to much simpler food, including hot dogs and hamburgers, for the day. But we managed to buy vast quantities of bread by going to the most obvious place. You didn't need any experience in the food industry to know that Sam's Club is a great supplier for small businesses like restaurants, and they never let us down. We rapidly became their single best customer in Puerto Rico, and they treated us well: we skipped the lines to enter through a side door. They were our lifeline when we began to exhaust supplies at José Santiago, and they needed time to restock. We always adapted our cooking to whatever ingredients were available. I could never understand the other NGOs struggling with supplies when the answer was as easy as Sam's Club. We had one Jeep and we filled every corner of it, from floor to roof, with loaf after loaf of bread.

But we were running late that day because of the move, and the result was a huge problem. The food trucks missed the lunchtime window for handing out hamburgers and didn't know what to do with the seven hundred left behind. To me, the solution was simple: go to where you know there are lots of people. "Stop at a gas station and give them away," I said. Within two minutes, all seven hundred burgers were gone.

I needed people who could solve problems like that, and fortunately one arrived that day. Erin Schrode, an activist who had worked in medical relief after the Haiti earthquake, came to help, alongside Nate. She arrived just as my executive director, Brian MacNair, was leaving to keep our operations going in Nicaragua, where he was traveling on a training trip with two of my chefs. I really needed all the help I could get, but we couldn't leave our other programs behind. My private business and family life were already suffering enough. Brian could help me by maintaining our other programs and supporting us from a distance.

"We're going to serve 100,000 meals a day," I told Erin. She looked at me like I was crazy. That day, because of the move, we were only going to prepare eight thousand.

Erin wanted to know what her responsibilities would be, but I didn't want to play the title game. When you give people titles, they screw up everything. They need to figure out how to make things work, not argue about their titles.

"You are chief of happiness," I told her, hoping to end the conversation.

"Okay," she replied. "I'm chief of operations and happiness."

There wasn't much happiness on the rest of the island. The official numbers were getting worse, not better. Héctor Pesquera, the island's secretary of public safety, admitted to a reporter that the number of dead exceeded the official count of sixteen, but said he didn't know how many. "I understand that there are more dead here, but what I do not have is reports that they tell me [for example] in Mayagüez eight died because they did not have oxygen, in San Pablo four died because they did not receive dialysis," he said.[1] I had seen four dead bodies myself, so I thought his numbers were far too low.

The situation on the island had hardly improved much. Governor Rosselló reported that only two-thirds of supermarkets were open, and in my experience, many of the open stores had no supplies to speak of. People had no money, and if they did, their cards weren't accepted in the supermarkets because of a lack of internet or power. Only slightly more than a third of islanders had cell phone service and, according to our volunteers and delivery drivers, most of those seemed to be near San Juan.[2]

For the people of Puerto Rico, we needed to grow quickly. But we also needed to grow at El Choli in a much more organized way. FEMA was now talking about a contract to get us going, which would require a whole new level of paperwork that might be audited at a later date. It wasn't yet clear how far that contract would take us. FEMA seemed to be taking our work seriously, not least because of the attention we were getting in the traditional media and on social media.

However, that media attention became vastly harder to attract because of the horrific events in Las Vegas a few hours before we started cooking at the arena. A gunman had opened fire on a country-music festival near the Vegas strip, leaving fifty-eight people dead and 851 injured. The numbers were stunning, and the carnage was shocking, even after the recent news of so many mass shootings. You could feel the news oxygen getting sucked out of Puerto Rico, just as we were finally gaining some traction with reporters from the mainland. Anderson Cooper could no longer stick around for days on end, as he had done in Haiti. He and the rest of the mainland media needed to fly to Las Vegas as quickly as possible.

I knew these were difficult news decisions to make. It was impossible to compare one tragedy to another. But it should have been possible to cover both disasters at the same time. One was quick and bloody. The other was slow-moving but killed many more. Both deserved extended attention from journalists.

The result was a shutdown of the coverage of Puerto Rico's suffering. The island became once again the forgotten disaster, where American lives did not seem as valuable as those in places lucky enough to hold full statehood in the United States of America. The loss of news attention meant that so many problems that could have been solved, and so many scandals that could have been exposed, escaped attention for many months. The government officials at all levels—from the mayors to the island government to the Trump White House—dodged any meaningful scrutiny. The tragedy of Las Vegas was also another disaster for Puerto Rico.

That night I returned exhausted to the AC Hotel, after spending a couple of hours at FEMA getting my hands on my first military-drawn map of the island. It arrived just in time to explain my plans to the island's secretary of education, Julia Keleher. With more than a thousand schools under her control, Keleher was a powerful leader in what I believed could be the most dramatic expansion of all the feeding operations. My visit to the school in Ponce was proof of what they could achieve. Every school has a kitchen and its own team of cooks, and it would only take an order to activate those kitchens—to get them to cook longer and with more supplies—to feed their own communities. School kitchens could be a model for how to feed an island, if we could just organize and activate them.

The schools were suffering in their own right. Most were closed, and communications were hard. But some were designated as community shelters and a few were already feeding families and relief workers. The key was to get the message to more schools—which was a great challenge without cell phone service, landlines and the Internet—and to get more food to them so they might serve their communities. The schools' kitchen staff needed to know they would not get into trouble if they started cooking more meals: not just for the students but also for families who were going hungry. As we looked over my new map, and talked about how to serve the communities in the greatest need across the island, I felt like Keleher could be one of the best partners I could hope for. In my eyes, she was so much more than an education secretary: she controlled the largest number of kitchens on the island.

The map was so much more than a piece of paper. You know what the map really became for me? A way to show that I wasn't crazy. The map meant I wasn't just a crazy chef who wanted to open eighteen kitchens. I could walk people through every step of the plan, through everything we had built and delivered, through every region and town that needed help. I was showing that all those activations of kitchens had a footprint. They were proof of what was possible, and I hoped that proof would help FEMA and the Red Cross to help other people. They might even help me feed those people. Thanks to my friends in the Army Corps of Engineers, especially Andrew Goldblatt, I even had a digital version of the map. It could have been so useful for everyone responding to the crisis, as an updated picture of all the resources and needs on the island. In my brain, this was a prototype for future disasters. If only people could get behind this project and what we were doing. Instead, I didn't have FEMA credentials and could only get a temporary password to see the digital map.

If they couldn't help or wouldn't help, I would call everybody who could help, who wanted to help. America is full of money, so I kept moving forward, with my phone and my maps. It was like opening a restaurant without knowing who your investors are. The return on investment was knowing that Puerto Ricans would not go hungry.

THE SCENE OUTSIDE EL CHOLI WAS A MESSY, NOISY, WONDERFUL COMBI-nation of ingredients. The constant buzzing of generators was the backbeat to the scraping of huge paella pans outside the west entrance to the giant arena where we'd set up our headquarters. As we organized ourselves with giant flip-chart sheets, and letter-size sign-up sheets, the cooks were heaping up mountains of rice and sloshing together large vats of stock. The hot meals would take shape just a few feet away from a crush of sweaty volunteers and anxious Puerto Ricans waiting to pick up the precious trays and boxes of food for their communities. The demand was so high that people arrived before dawn—some as early as four in the morning—waiting for hours to collect meals for their community. We hired extra security to manage the crowds, organizing them into lines for placing orders and picking up food. Close to the paella pans, our small fleet of food trucks waited to be filled with what would become several lifelines stretching across the island each morning. Our outdoor kitchen was also our storage depot: there was a trolley of propane tanks, ready to switch into the circular burners that were firing up the rice. By the arena entrance, a half-paved car park was always full of trucks and cars waiting to take our food across the island. There were heavily armed Homeland Security agents, with assault rifles over their shoulders and sidearms strapped to their legs. And there were pastors and mayors from small communities with young volunteers, all patiently waiting in the humid sunshine. When it rained, which it often did, the car park would turn into a soaking mud bath, just like much of the rest of the island, which was still saturated with groundwater after two hurricanes.

The evidence of the storms was all around us. On one side, the home of the Puerto Rico Trade and Export Company bore the scars of the winds: its logos ripped off the building, leaving empty white space which once boasted of its business. On another side, an office block was boarded up with several plywood sheets, its tinted windows blown out. There was no sign of repairs going on in our neighborhood.

Inside the arena doors we started storing huge supplies of our basics: cutlery, paper plates, clamshell boxes, aluminum trays, pallets of water, cans of vegetables, trays of bread, boxes of fruit and vegetables. In normal times, not so long ago, this was the Absolut Vodka lounge, and the walls were covered in playful takes on the vodka lifestyle. There were cocktail glasses and playing cards alongside giant warnings to drink responsibly. In this darkened lounge, just beyond the tropical sunlight, we set up two new sandwich lines on foldable tables. The space was big enough to take hundreds of volunteers and could be a factory for feeding the island. Beyond the sandwich factory was the half-empty floor of the arena, where the first lady's supplies stood still, unused for reasons that were not at all clear to us. We were forbidden from entering the space, so we could only look at the stack of new generators and wonder why they weren't getting to the people who needed them so urgently.

Around the corner, through the long cinder-block hallway, we stored more ingredients: rice, oil, cans of vegetables, boxes of stock cubes, tray after tray of white sliced bread and giant bottles of canola oil. Beyond another lounge area that also opened up to the main arena floor, there were more supplies in the hallway that led to our central kitchen: a giant space that normally churned out arena food. On one side we were cooking vats of chicken and rice, *pastelónes,* even hot dogs. Behind a wall, we were washing and prepping piles of vegetables and raw chicken. And just behind the prep area, there were two giant walk-in refrigerators that were critical to the whole operation.

Our cooks soon filled the walls with ripped cards where they wrote their targets for the day: the numbers of meals that would go to each location, and most important of all, the number of hungry Puerto Ricans they were feeding each day.

Compared to our first home, El Choli was a five-star facility. More than just its size, the space allowed us to be efficient in our cooking. We had water and heat, the two essentials of cooking that were still only a memory for most Puerto Ricans. We had refrigeration and air-conditioning, even if the heat of the kitchens was overwhelming. We had space to deliver mass quantities of food, and space to store them too. We could finally set about feeding the huge numbers of people who were going hungry while they tried to rebuild their lives.

Outside, on more foldable tables, I spread out my new maps and planned our expansion across the island. With this giant kitchen and outdoor space as our central hub, we could look after San Juan. But to get to the rest of the island, we needed a network of kitchens doing the same thing as El Choli but on a smaller scale. From those satellite kitchens, we could serve the local communities and figure out where the real needs were. We needed to activate those kitchens and get supplies to them. In doing so, we would avoid the difficulties of daily travel with hot food, improve our intelligence and communications about the facts on the ground, and expand to be truly island-wide. This was our twenty-one-day plan to feed the island. FEMA had its own twenty-one-day plan, but they could never say when the clock on the twenty-one days began.

On the walls we stuck photos of the scenes in Santurce as a

way to show with pride what we already achieved. They were a message of hope about what we could do together. Anybody who saw the photos felt inspired to join our movement.

What gave me confidence to grow was the first official sign of support, just eight days after I landed on the island. FEMA emailed to give us what they called "a notice to proceed" to cook twenty thousand meals a day for the next seven days. We quoted a price of $6 to $8 a meal, depending on the ingredients, covering not just the food and supplies, but the transport and power required to make and deliver it. For reasons that were never clear to me, that cost per meal was bumped up to $10 by the time FEMA gave us the go-ahead. All told, we were heading toward a FEMA contract worth $1.4 million to provide 140,000 meals. We knew we could hit those numbers. On our last day in Santurce we had produced twenty thousand meals. Now we were in a vastly bigger space, with new kitchens ready to activate across the island. Our plan was to cook to capacity, not to the contract. As a nonprofit, we had no interest in making a profit. Our sense of profit came from feeding the people. FEMA's money was going to cover our current costs and beyond: anything left over we would spend on feeding more people, regardless of the number of meals in the contract.

PERHAPS IT WAS JUST A COINCIDENCE THAT FEMA'S EMAIL LANDED ON the same day that President Trump was visiting Puerto Rico for the first time since the hurricane, fully thirteen days after Maria made landfall. But I wanted him and his team to know what we were doing, and what we could do together. My voice was turning scratchy and my throat was painful, but I recorded a video to give him—or his aides—an idea of what we could achieve.

"Hi, Mr. President Trump. This is how we're going to be feeding Puerto Rico for the next twenty-one days," I began, standing next to the maps on the tables outside El Choli. "There's a lot of things happening, with the Red Cross and other relief organizations. But this is World Central Kitchen and this is what we're doing quickly. We are here in our headquarters where we're already producing 50,000 meals. By Saturday we'll be doing 100,000. We identified eight kitchens around Puerto Rico that we're going to be opening: Mayagüez, Aguadilla. We're going to do Manatí, and we'll open one in Ponce tomorrow. We'll do one in Guayama, and one in Fajardo to take care of Vieques and Culebra. And we'll do one in Gurabo to take care of the center of the island. We are going to do in a week—if we get direct support, we are working with FEMA to make sure we're actively feeding with this project—roughly over half a million people a day. Plus the other operations that are already happening, I think, Mr. President, we will feed the island. We only need to make sure that we put away the red tape and we keep moving forward. We have over 200 volunteers a day here. We can be doing this. We only need leadership."

Leadership was all we needed, but Trump's visit did not exactly demonstrate the leadership we were hoping for. White House staffers were worried that there would be a repeat of Trump's angry outbursts against Mayor Cruz.[3] Instead of playing the traditional role of a president comforting American citizens in crisis, Trump might turn back into the spiteful tweeting politician the world saw just a few days earlier. His aides were concerned about protests, and critical comments from officials like Cruz. They might even have been concerned about someone like me telling the truth about the disaster of the recovery operations.

They had plenty of reasons to worry about their boss. The whole trip was designed to shower praise on Trump and for Trump to return that praise to his own team. He didn't venture outside of his comfort zone, which was very small indeed. Even before he left Washington, Trump told reporters that the recovery was awesome. "I think it's now acknowledged what a great job we've done," he said at the White House.

The only person doing the acknowledgment was Trump himself. At a fake briefing in an aircraft hangar at the Air National Guard base, on the site of the main San Juan airport, Trump began strangely by praising the weather in Puerto Rico. Soon he was praising Brock Long, his FEMA administrator, in ways that echoed George W. Bush after Hurricane Katrina, when he said his FEMA chief had done a heckuva job.

"Brock has been unbelievable," Trump said, after giving him an A+ for his work in Texas and Florida. "And this has been the toughest one. This has been a Category Five, which few people have ever even heard of—a Category Five hitting land. But it hit land—and boy, did it hit land." He continued to praise his team before turning to the governor.

"Governor, I just want to tell you that, right from the beginning, this governor did not play politics. He didn't play it at all. He was saying it like it was, and he was giving us the highest grades."[4]

If it sounded like an exercise in giving good grades, that's because it was. The person who seemed to need the most relief was Trump himself. When he started to talk about the suffering in Puerto Rico, it was detached from any sense of human feeling.

"Now, I hate to tell you, Puerto Rico, but you've thrown our budget a little out of whack because we've spent a lot of money on Puerto Rico, and that's fine," he explained.

"We've saved a lot of lives. If you look at the—every death is a horror. But if you look at a real catastrophe like Katrina, and you look at the tremendous hundreds and hundreds and hundreds of people that died, and you look at what happened here with, really, a storm that was just totally overpowering—nobody has ever seen anything like this. What is your death count as of this moment? Seventeen? Sixteen people certified. Sixteen people versus in the thousands. You can be very proud of all your people, all of our people working together. Sixteen versus literally thousands of people. You can be very proud. Everybody around this table and everybody watching can really be very proud of what's taken place in Puerto Rico."

Never mind that the official death count was obviously wrong, or that there were credible reports of the morgues being full across the island. What kind of person is proud of sixteen people dying? Only someone who could draw a line between "all your people" and "all of our people." Someone who didn't see Puerto Ricans as "our people" or real people at all, even though he was the president of all American citizens, including those in the Caribbean. We needed leadership to feed the people of Puerto Rico, but we had a leader who was more interested in patting himself on the back than the difficult work of disaster recovery.

The day would get worse with Donald Trump. He drove the short distance to Guaynabo, where a church called Calvary Chapel was using an old Office Depot warehouse to store and distribute hurricane relief supplies. In front of a table stacked with rolls of kitchen paper and cans of food, Trump looked at the excited crowd of Puerto Ricans taking his photo and started lobbing the paper rolls at them. Like he was shooting a basketball at an imaginary hoop, he turned to one side and threw a roll, then turned and shot another.

He seemed to have no idea what his role was, as president of a country in the midst of a humanitarian disaster. At one point, he picked up a water purification packet and seemed astonished that anyone would use it.

"Wait, you put it in dirty water?" he asked.[5]

"Yes, and you can drink it in ten to 12 hours," said one volunteer.

"Would you drink it?" he asked, sounding more incredulous with each word.

"Sure," said the volunteer.

"Really?" he asked again, looking alarmed.

"Really," she replied.

Trump looked at the purification packet, held it at arm's length and scowled. A famously germophobic man, he looked like he'd rather drink a bottle of Purell. He handed out some cans of food, posed for a few selfies, then turned to the cameras behind him.

"There's a lot of love in this room," he declared. "A lot of love in this room. Great people."

He meant: there's a lot of love for me.

His last stop of the day was with the governors of Puerto Rico and the U.S. Virgin Islands, on board the USS *Kearsarge*. Technically the massive ship is an amphibious assault ship, but in practice it has often been deployed for humanitarian missions, including the evacuation of some three thousand civilians from Sierra Leone in 1997, as its capital, Freetown, descended into violent anarchy. Its medical operations are only second in size to the hospital ships *Mercy* and *Comfort*, which also arrived in San Juan that day. The ship has such massive capabilities that it can distill 200,000 gallons of water a day: enough daily fresh water for the entire population of San Juan.

Sailors and marines from the *Kearsarge* were on the ground the morning after María passed through.[6] With *Kearsarge* in Puerto Rico, the United States had every capability needed to minimize civilian deaths in the most extreme circumstances. The ship's helicopters were flying relief supplies into the island, as well as helping with urgent repairs to equipment like hospital generators.[7] But five days after the hurricane, they had only made eight medical evacuations.[8] Compared to its service in Sierra Leone, it seemed like the *Kearsarge* wasn't being used at anything like its full capacity. Now it was simply a stage for a photo op, with the Virgin Islands' governor arriving on board to help Trump avoid the trouble of flying to his devastated island.

Trump was pleased with his day in the disaster zone. On the way back to Washington, aboard Air Force One, he interrupted a media briefing by Jenniffer González-Colón, the island's non-voting representative in Congress. One reporter asked Trump if he'd heard any "constructive criticism" on his trip.

"None," Trump said. "They were so thankful for what we've done." At that point, Trump noticed the reporters included Geraldo Rivera from Fox News. He said hello, even though Rivera had just interviewed him on the ground.

Trump rashly told Rivera that Puerto Rico's crushing debts would be canceled by the United States. "We have to look at their whole debt structure," he said. "You know, they owe a lot of money to your friends on Wall Street, and we're going to have to wipe that out. You can say goodbye to that. I don't know if it's Goldman Sachs, but whoever it is, you can wave goodbye to that."[9] This was actually an awesome idea if it ever became something more than an off-the-cuff opinion.

One reporter asked Trump what his visit meant to Puerto Rico, and the president was happy to oblige. "I think it means a lot to the people of Puerto Rico that I was there. They've really responded very nicely, and I think it meant a lot to the people of Puerto Rico. I mean, I think you folks have seen it. And I guess it's one of the few times anybody has done this. I didn't know that at the time, but I guess, from what I'm hearing, it's the first time that a sitting president has done something like this."[10]

Sitting presidents make visits to hurricane-hit parts of the United States all the time. And sitting presidents have visited Puerto Rico many times before, including Barack Obama in 2011. It wasn't clear what Trump was talking about, or whether he understood what was going on in Puerto Rico.

Back on the island, the reaction to his visit—especially the paper towels—was one of dismay. Even by Puerto Rico's low expectations, Trump had managed to disappoint everyone I met. He lobbed supplies at Puerto Ricans rather than taking seriously and sympathetically the humanitarian crisis in front of him.

"Forget politics, forget pundits," I tweeted, trying to turn the conversation back to the positive work I was seeing firsthand. "What I have seen in #PuertoRico is people coming together, sacrificing 2 serve. This is humanity at its best."

THE TRUMP ADMINISTRATION WAS SECRETLY DOING MUCH MORE TO deliver food relief than it ever admitted to me or anyone else. Which made it even worse that the food relief never materialized.

On the day Trump was tossing paper towels into a crowd of Puerto Ricans, his own FEMA officials signed a contract worth almost $156 million with a tiny contractor based in Atlanta to provide 30 million meals. The company, called Tribute Contracting, had one employee and no meaningful experience in food production. Just twenty days later, FEMA canceled the contract after Tribute produced only fifty thousand meals.

Tiffany Brown, the owner and sole employee of Tribute, said she learned of the FEMA bid from a Google alert. She landed the contract by bidding at $5.10 a meal, and planned to subcontract the 30 million meals to two small caterers. One of them, a wedding caterer, had eleven employees. She insisted that the subcontractor "was experienced with this work" and would hire more people as they scaled up.[11] They were going to freeze-dry mushrooms and rice, chicken and rice, and vegetable soup. Another contractor, a Texas nonprofit, was going to ship the food to Puerto Rico. "My biggest mistake was not asking for more help," she said later.

In addition to not delivering the meals, Tribute failed a key FEMA test. The meals were supposed to be "self-heating" but the supposed heating pouches were shipped separately from the meals.

Of course, those FEMA tests are as stupid as they sound. Freeze-dried food needs water to be reconstituted, but clean water was in very short supply in Puerto Rico. And the idea of self-heating food could only come from a government official who has never actually cooked a meal that anyone wants to eat. Tiffany

Brown was not the problem; she was just a symptom. The sickness is a food relief system, managed by FEMA and the big nonprofits, that does not understand food or the people it's supposed to be feeding.

Brown describes herself as "a broker" taking a cut of the contracts, much like Josh Gill. Her expertise was in contracts, not food. "They probably should have gone with someone else, but I'm assuming they did not because this was the third hurricane," she said later. "They were trying to fill the orders the best they could."[12]

That's being kind to FEMA. Another company called Bronze Star won $30 million in contracts to provide half a million emergency tarps and sixty thousand plastic sheets to cover roofs damaged by the hurricane. From what I saw across the island, Puerto Ricans desperately needed the tarps to stop the rain flooding what was left of their homes. But the tarps never arrived because Bronze Star was just another tiny contractor: two brothers with no experience and no track record, relying on other suppliers who failed to deliver. The tarp contract represented one-third of all the FEMA money spent on tarps at the point when it was canceled. The brothers were both former military personnel, but neither had won a Bronze Star.[13]

FEMA said that everyone who needed a tarp could get one. And it said that nobody missed a meal because of Tribute's complete failure to provide anything edible in any quantity. "At the time of the contract termination there were ample commodity supplies in the pipeline and distribution was not affected," said William Booher, a FEMA spokesman.

At best, this is meaningless jargon: a pipeline is not a meal, and distribution was not affected because there was so little to distribute.

At worst, this is a shameless lie.

The contract only came to light when Democrats on the House

Oversight Committee demanded a subpoena of FEMA to turn over documents about the contract. Tiffany Brown told committee staffers that she worked hard to provide the meals, "24 hours a day, seven days a week." But her suppliers stopped working because FEMA was late with payment.[14]

How could FEMA know Tribute would fail? Maybe they could have checked its capacity to deliver such a huge order, never mind pay its suppliers. Maybe they could have checked the federal contractor database, which showed Tribute had failed to deliver on four other food contracts in 2013 and 2014. When Tribute failed to deliver a contract for tote bags, another federal agency—the Government Publishing Office—said the company should not get a contract worth more than $30,000 until 2019 at the earliest.[15]

On the day the Tribute contract was signed, we had already prepared and delivered 78,000 meals, most of which were cooked that day with healthy, fresh produce. The meals didn't need a self-heating pouch because they were already hot. They didn't need shipping because they were already on the island. Much of the money we spent went back into the local economy, through Puerto Rican suppliers and producers. That included bread baked on the island for, at that point, fifteen thousand sandwiches. The private sector was doing just fine.

The Tribute contract was being negotiated at the very same time as FEMA's head of mass care in Puerto Rico was telling me the agency couldn't move that quickly. It was finalized while the Red Cross was suggesting to me that Chefs For Puerto Rico was too small to handle this crisis.

"José, you don't understand the process," Waddy González had told me at FEMA's offices. "We can't do this as quickly as you want."

To this day, I don't understand FEMA's process. Of course they could have done this as quickly as I wanted. I was only talking about one million meals when they were about to sign a contract for 30 million meals. What they meant was that they preferred to deal with a one-person contractor in Atlanta rather than a group of chefs in Puerto Rico. They preferred a for-profit business 1,500 miles away over a nonprofit with good intentions and great understanding of the needs of the island.

People liked to say at the time that FEMA was overextended, and it's true that overall, during that same time period there were many hurricanes and many floods in many places. But the truth is that FEMA had the people and the money to do much better in Puerto Rico. They chose not to do the right thing for American citizens living in the worst conditions. The problem might have been too big for their brains, but the solution was right there in front of them. We didn't want a contract because our business was winning contracts. We wanted to feed the people.

FEMA weren't the only ones to give huge contracts that didn't pass the smell test. PREPA, the Puerto Rican public power company, gave a $300 million contract to a tiny Montana contractor called Whitefish Energy, which first contacted PREPA through the LinkedIn social network. One of Whitefish's main investors was a major donor to Trump's presidential campaign, and the CEO was friends with Trump's interior secretary, Ryan Zinke.[16] But the company had no experience or resources to do the job. With only two full-time employees, Whitefish hired workers from the mainland at $63 an hour, and billed PREPA five times that amount, not including travel, food and hotel costs. Those massive profits were in sharp contrast to the traditional way utilities help each other out after a natural disaster, sending workers at cost when the requests come in. PREPA had overruled its own lawyers in signing the Whitefish contract, after no competitive bidding process. The contract strangely stated that it could not be audited.[17] As Puerto Ricans continued to struggle with no electricity, Whitefish became shorthand for the scandal that nothing was getting fixed while somebody was getting rich. In the face of huge public criticism, the governor ordered the contract to be canceled after just a few weeks of work.[18]

Failures like these don't happen because of one person or one agency. They happen because the whole system breaks down. They happen because the president wants to play golf, or pat himself on the back or throw paper towels at his own people. They happen because his own administration is too proud or too stupid to find the right people to do the right job in a crisis. They happen because the local leaders are too busy playing local politics, or talking to TV cameras or hoping the president will solve all their problems. They happen because the big charities and nonprofits are more interested in preserving their power and their budgets than the expensive work of getting food, water and power to the people.

They certainly don't want a loud-mouthed chef telling them to think in a different way.

That day we doubled our meals, from eight thousand to sixteen thousand. El Choli was ramping up on our way to my dream of 100,000 meals a day. Orders were coming in and we never said no. "We have big plans because people have big needs," I told my team.

THE DAY AFTER TRUMP VISITED, FEMA OFFICIALS ASKED ME TO GO TO the government headquarters at the San Juan Convention Center first thing. The mass care team under González wanted to meet in person, and promised to escort me into the building. I still had no official credentials, and our FEMA contact was not around when we arrived. So we found our familiar side door and walked right in. I was waiting at the meeting space—a large hallway—when a security guard, armed with an automatic rifle, confronted me.

"Where is your pass?" he demanded.

"I don't have one," I explained. "I don't know why. They still haven't given me one."

"How did you get in?"

"Through the side door. You should check it," I told him, sarcastically trying to help.

At that point, González came over with his team. I started unfurling my map on the floor, explaining in detail my twenty-one-day plan to feed the island, while the armed guard stared at the whole scene. The man tasked with feeding the island looked at me like I was an alien, when all I was doing was giving him the plan he needed to do his job, to help our fellow Americans. We needed military support to provide fuel and water for all the kitchens we planned to open. If we all worked together, we really could feed the whole island. As I talked, more security guards arrived.

"You don't understand, José," González kept telling me.

I looked at him and wished we had a president in office who could make me the food tsar of Puerto Rico. You can achieve so much if you have the power to cut through the bureaucracy that makes the federal government so expensive and inefficient.

I felt the security guards bearing down on me, as I kneeled on the floor with my maps. Nate walked over but they yelled at him to stay back, and the grip on their guns tightened. They insisted I get up and leave, so I did, under four armed escorts.

"You should check out that side door," I reminded them as I walked out. "And actually there's another one." They thanked me as we traveled down the escalator. Some FEMA staff recognized me as we left and thanked me for my work. At least we were leaving through the front door, I thought.

On my way out, I met with the real experts in food relief: the Southern Baptist Convention, whose mobile kitchens are so critical in these disasters. However, they had no capacity to transport their mobile cooktops to Puerto Rico. I suggested I could activate a kitchen for them, complete with volunteers, but it was all too complicated. Instead, they were housing one hundred homeless people and doing some door-to-door searches to reunite families. "It's too complex," Jack Noble of the SBC told me. If the Southern Baptists weren't cooking, then there was no way FEMA, the Red Cross or the Salvation Army could feed the island. There was no other group who had the ability or the experience to do the job. Other than us. I knew it because I had watched them at work. Instead of cooking, they were focused on rebuilding homes and handing out tarps to stop leaking roofs.

"We decided we are going out alone," Noble told me. "It's time for the Southern Baptists to step up and mobilize our own volunteers. It's not everybody else's job to make us look good. We have to step up and do our own work. It's our calling. It's our job. It's our mission. We need to help people think about how they are going to get along without electricity for nine months. This is America's biggest disaster in its history and that's the bottom line. The fact that this is an island just complicates the matter so much. If this was on the mainland, it would be significant but the fact that this is an island just compounds the difficulty of the response. It even makes partnership more difficult because it makes you a little bit hesitant about sharing."

I admired the SBC enormously. They were serious about their work and felt a responsibility to help people. But he was right about the scale of the disaster—the biggest in living memory— and the challenges of being on an island. This wasn't Southern Baptist country and they had no reason to overcome the logistical hurdles of being on the island. They also had no real support or leadership from the Red Cross, their normal partner, to solve these problems. Chefs For Puerto Rico was entirely different: our chefs and kitchens were already on the island, and we knew the suppliers whose very businesses were built on overcoming the logistical hurdles of being here.

I returned to El Choli believing the worst about FEMA. Yes, we had our first contract in the works, but it wasn't finalized and the money seemed a long way off. More important, they didn't care for my plan to feed the island, and weren't interested enough in my work to give me a security badge or tell the security guards to stand down. I gave an interview to Jorge Ramos of Univision and I could barely hold back my tears in front of the TV cameras.

OUR OWN FOOD OPERATION WAS DOING SUCH GOOD WORK, IT WAS HARD to miss the spirit of what we were creating. We were determined to set a record for the number of meals we prepared that day, with the sandwiches and the hot meals. Our food trucks fanned out across the island and what they reported back lifted us all.

Xoimar Manning, in the Yummy Dumpling truck, said she found some community workers when she arrived as usual in Loíza, a small, poor community on the eastern tip of the island, where Maria first ripped through.

"They were talking to each other, talking about food," she said. "They saw my T-shirt and immediately recognized World Central Kitchen. They said they went to El Choli today but they only got food for two hundred people and they were super-worried because the community is much bigger. I told them we had come with 1,500 meals and they started to hug me. I cried."

I began crying too when I heard her story. Every day stories like this popped up in my WhatsApp chats, or were scribbled on a piece of paper that was handed to me, or were told to me in person by someone who had waited for hours to see me. I have always looked like a strong boy and tried to protect that image. But I was astonished by these stories and heartbroken for the many people still suffering without clean water and food.

Around the same time as the food trucks reported back, FEMA seemed to reverse itself. With a single tweet, they appeared to endorse my twenty-one-day plan, posting a video of me explaining the plan and talking about my partnership with agencies including FEMA itself.

"Here's @chefJoséandres sharing his plan to feed #Maria survivors across Puerto Rico with support from federal partners," FEMA tweeted. After all the struggles and confused stares from FEMA officials and the big charities, it felt like vindication.

"Team, you can be proud!" I told my chefs. "FEMA has just announced that our plan is the plan to follow!"

Later on, I realized I was being naïve. They were just using my image like Katniss Everdeen is used in *The Hunger Games* propaganda. It was a rollercoaster of despair and happiness. That was Puerto Rico. One moment, you were freaking out; the next, you were thrilled. For a brief time, I felt like I belonged, like they recognized their shortfalls, like our plan could be successful.

The team could also be proud of their performance that day. We set a new daily record with 21,965 meals prepared: more than we managed on our last full day in Santurce. If that weren't enough of a milestone, we also crossed 100,000 meals in total since we'd started cooking, little more than a week ago.

Water was now our biggest challenge because there was such a shortage across the island. But we were making steady progress on water too. One friend promised to donate eleven pallets of water, which would be eighteen thousand bottles. We bought eight thousand bottles from my friend Alberto de la Cruz at Coca-Cola, at the relatively low cost of thirteen cents a bottle. But we were all frustrated by the water crisis. De la Cruz was selling his water to FEMA, and had very little to spare for us. Pepsi's production was offline. FEMA's supplies of water were getting distributed by the Red Cross, and they refused to share them with us. I knew there were places where there were stockpiles of water when there were many thirsty people on the island. When we got our deliveries of food, it all disappeared within hours. Why was water different? Sourcing more water was a daily struggle that often ended in failure. If you got your hands on any water, the task of trucking the bottles was no easy feat.

"There are 120 water oases in the island with more than enough water for everyone, but the director of health is telling people the water is no good, so nobody wants to use it," de la Cruz said. "So you have the ridiculous situation of people buying water to wash their children." When the person who makes a living from selling bottled water tells you the situation is ridiculous, you know the system is truly broken.

In fact, the island's department of health was telling people the water was so polluted they shouldn't walk in it either, because of animal urine. The levels of fear around water were far above the real threat to public health, and the health department was struggling to even complete its water tests because the testing labs were damaged by the hurricane.[19] It didn't help that there were reports of people drinking from wells at Superfund cleanup sites, and the Environmental Protection Agency issued a blanket warning against such wells.[20] The EPA later found that the wells in question were safe for drinking but nobody believed the official advice by that point.[21] Bottles were all they trusted, and bottles were expensive and in short supply.

"FEMA wants to buy my water but they want to buy only small bottles," Alberto said. "I said I didn't want to sell them small bottles because people will throw them away. I wanted to sell the gallon bottles to go to the oasis to fill them up there. But they created this panic and now there isn't enough bottled water. It's like the diesel situation. We didn't have a supply problem; we had a distribution problem. We tried to resolve the water issue but we couldn't get the department of health to agree with the water authority."

I believe that the private sector could have solved the water challenge in no time, just as it solved the gas and diesel challenge early on, with Alberto's help. His bottling plant was so well prepared for the hurricane that they managed to be up and running again within two days of Maria. They placed tarps on their machines so the water didn't ruin them, and they worked closely with the local mayor to clear the roads from the plant.

Back in Washington, the Trump administration was already backing away from all the happy talk of the president's visit. Mick Mulvaney, Trump's budget chief, dismissed Trump's strong suggestion that the administration would help solve the island's crushing debt problem. "I wouldn't take it word for word with that," Mulvaney told CNN. "Puerto Rico is going to have to figure out how to fix the errors that it's made for the last generation on its own finances."[22] I still thought we needed a Marshall Plan for the island. They could have called it the Trump Plan if it helped make it a reality for the island to recover and prosper.

Reality was starting to bite. At a press conference, Governor Rosselló admitted that the death toll was higher than what Trump had bizarrely bragged about the day before. Instead of sixteen dead, Rosselló said the total was thirty-four. But his own official at the press conference suggested even that number was wrong. "I don't think this will be the final number," said Héctor Pesquera, the island's secretary of public safety. "And we've never said it will be."

Pesquera said they'd been slow to update the death count because the bodies were stuck at hospitals that were out of contact with the island's government. "The bodies weren't coming in," he said. "There was no way of transporting them. They were in the hospital morgues and there was no communication with hospitals."[23] If I were in their position, I would send teams around the island to check on their operations just as we were checking on our kitchens.

Whatever food we were preparing and delivering, it was the least we could do for an island whose suffering was so hard to calculate.

TO FEED AN ISLAND, YOU NEED TO THINK ABOUT THE WHOLE GEOGRAPHY. Even with a big headquarters like El Choli, there was only so much we could produce and distribute from San Juan. By spreading out across the island, we could cook so much more food. But more important, we could pick up precious intelligence about what was going on. Communications remained patchy outside San Juan, and communities had no way of knowing—beyond local radio news reports or social media—that we were cooking. Puerto Rico prospered and suffered from its island culture: everyone seemed to know everyone else, yet it was hard to separate rumors from reality. For real information about the rest of the island, we needed to leave San Juan. We certainly needed to leave the government headquarters at the convention center.

Our plan, as we pinpointed on my precious maps, was to open kitchens across the island. For that, I needed a partner with lots of kitchens. We were already working with the Department of Education to get school kitchens to cook for their communities, but it was hard to know if those schools were doing the job. Even Secretary of Education Julia Keleher had poor contact with her own schools. The potential was real. All they needed to do was to find the local leaders and get their food to the elderly in the *egidas* and the homeless in the *refugios*. So we recorded a video together at El Choli in which I urged the schools to open their kitchens to meet more of the communities' food needs. She posted the video to her social media accounts, saying this was a call to school kitchen employees to work together to feed those in need. But we had no idea if her staff would see or hear the message.

Still, there was a model there. Schools are a tremendous resource in a time of crisis, and there are many different types of schools, not just for children. I have worked a lot with culinary schools on the mainland, and I know there are few places with more professional kitchens and cooks with basic training. I started talking to the vocational school network called the Instituto de Banca y Comercio, or IBC. If the Southern Baptists weren't coming with their mobile kitchens, perhaps we could replicate their operation with a network of regional kitchens. IBC was a group of schools that were only loosely managed from the center, but their enthusiasm and their facilities were unmatched. They could prepare the meals, at a cost, and with the support of our deliveries. These were the kind of suppliers that a FEMA contractor needed to find: local partners, with technical expertise and local facilities. They could produce what we wanted, under our supervision, in real time and at large volumes. Between the IBC kitchens, the regular schools, my own kitchens and a commercial caterer I had found, we could feed as many as one million people every day. I don't want to sound crazy but crazy is sometimes what you need to look for.

We started our partnership in Ponce, in the south, serving the second biggest city on the island. My promise to Ponce's mayor was important to me. It was the beginning of an important transition for us. We needed to show that our model could be repeated time and again, for Puerto Rico and for food relief operations anywhere in the world. Our expertise was not just in cooking, and we couldn't be the only ones to cook the food if we truly wanted to feed the island. Our expertise was in the whole food chain: from understanding what people wanted, to establishing where hungry people could find the food; from securing reliable supplies of ingredients, to distributing that food to the kitchens. We were matching supplies with needs, on an island where power and communications were still very unreliable. We had no idea how anyone had done this before, or how the official powers were planning to do it now. But we solved the problems as they popped up, as chefs do, and we just started cooking. Ponce was a case in point: the IBC kitchen needed our help to distribute their meals on the first day we visited. So my friend and board member Javier Garcia walked into the nearest church and asked the whole community to spread the word about our hot meals. Soon, those meals were gone.

With FEMA's apparent endorsement, and the arrival of military reinforcements on the island, I started imagining what we could achieve if we combined forces: World Central Kitchen with other NGOs and the U.S. military. I heard that Oxfam was coming out, and I started tweeting at Oxfam officials, urging them to contact me. If the U.S. Navy had a hospital ship, why couldn't we use a ship as a floating kitchen? We could cover the island that way, using the ship-to-shore transport normally used in military operations. With all the freshwater production on navy ships, we only needed to figure out how to get enough tankers filled every day and position them in the streets where people lived. We didn't need to bother with all those plastic bottles that were so hard to source and had ended up as a vast mountain of trash in Haiti. We could be creative with the massive resources of the U.S. government and avoid both a humanitarian *and* environmental disaster. My mind was dreaming of ways we could ease this crisis and every other disaster yet to come. If only someone could see what we were doing, how we were improvising. If only someone could see how people were suffering, and react with the same urgency we were. If only someone would listen to me.

"We only have 6 water tankers. We need 12 more," I tweeted at the state department and the defense department. "So we transport water faster and quicker. I buy them, you bring them?" Maybe we needed 100 tankers, but this was a start.

I never heard back on this or any other idea for how the military could help at scale. But that didn't stop me. People told me Twitter wasn't the way to do this kind of work. Oh really? With poor phone service, social media was good enough for anything. If President Trump can negotiate by Twitter, I am entitled to use the same platform. I started targeting the people who surely would be heard: the talented performers whose fame gave them a platform to talk about Puerto Rico. A plane arrived that day with several stars who wanted to help. I wanted to meet them at the airport but arrived too late for their press conference. I invited them to see our cooking at El Choli but their day was too busy already. Still, they brought a lot of happiness to many suffering people.

Luis Fonsi, the man who brought the sound of San Juan to the world with that summer's blockbuster "Despacito," arrived to hand out water bottles in La Perla. Fonsi was literally giving back to the historic and colorful slum in Old San Juan, where he made the video for "Despacito." I asked him on Twitter if he needed any help. He didn't reply but Lin-Manuel Miranda, the genius who created Hamilton, did. Miranda's Puerto Rican roots were strong enough to take him back for a month each summer to his grandparents in Vega Alta, west of San Juan.

"You have been so inspiring in this time," he tweeted at me. "Gracias." His support meant more to me than any FEMA video, not least because it would reach so many more people who might possibly help the hungry people of Puerto Rico.

Other stars were even more direct: Emilio and Gloria Estefan gave me envelopes full of cash to distribute around the island any time I encountered people in need.

That day we smashed through our previous record, preparing 25,828 meals. Just a week earlier, we'd produced less than one-fifth of that number.

BY THE END OF THE WEEK WE COULD SEE THE IMPACT OF SPEAKING TO the outside world, beyond the island. Donations were showing up at El Choli in significant volumes, although we were cooking so much food that the donations barely made it through one day's worth of meals. There were limits on what local businesses could donate because they were still struggling to survive. Meanwhile, donations from the mainland faced logistical obstacles in reaching us. Goya Foods led the way with a big delivery of juice, rice, yucca and other items. Goya, founded by Spanish immigrants in Puerto Rico, is the biggest Latino-owned food company in the United

States. Mario Pagán, one of the island's best chefs, introduced me to one of his childhood friends, Jorge Unanue, a Goya executive and part of the family owners, and the partnership quickly felt deep and personal. Goya shares so much of my view of the world: bringing high-quality food that people want, spreading Spanish and Latino culture across the world, and giving back to the community whenever possible.

They were by no means the only donors to deliver. UPS, which quickly established reliable deliveries to the island, brought us ten thousand bottles of water, which we sorely needed. The water tasted even better because some of it was donated by the New York Mets. Walmart helped us financially, while Chili's restaurants—run by my friend Ramón Leal—donated thousands of pounds of chicken that would be essential for our chicken and rice meals. The pallets and boxes began to pile up in the giant hallways and loading bays at the arena.

We began to think about how to sustain our funding, since FEMA was hardly prompt in payment and we still did not have a signed contract. We started asking for small donations on our World Central Kitchen website, and I asked my wealthy friends for help. I got my own restaurants on the mainland, including Bazaar in Beverly Hills, to donate food to local fund-raisers for Puerto Rico. Every donation helped raise funds, and awareness of what needed to happen.

However, the most important donation was the time and passion of our volunteers. We couldn't just rely on the chefs in Puerto Rico, because they were small, independent entrepreneurs like me. They needed to get their own restaurants back into business. We needed regular people to help with our massive sandwich operation. Some of them were hurricane victims themselves, homeless and hungry, and the meal they ate with us was their only meal of the day.

We also needed experienced cooks—people who knew how

to cook at volume—to run our hot meals. I flew over a team of my own chefs from my restaurants, led by David Thomas, my executive chef at Bazaar Meat in Las Vegas. David quickly helped bring some order and procedure to our work at the arena kitchen, to better manage inventory and all our culinary operations. He was central to building our capacity to grow so quickly. Chefs like David are problem-solvers, and that makes them ideal for figuring out how to meet the daily challenges of a food relief operation.

I also asked my friends running big catering and restaurant businesses back on the mainland if they could spare anyone, and the offers of help came back even faster than I could have hoped. Gary Green, who runs the Compass Group in North America, jumped into action, just as he did in helping me prepare 25,000 meals in Houston. After one email to my friend Fedele Bauccio, the CEO of the Bon Appétit catering company, I was introduced to Karla Hoyos, a twenty-nine-year-old executive chef who was at the heart of how we fed an island. Well before the hurricane, Karla was an activist: the president of an organization helping immigrant families in the U.S. dealing with deportation, or the criminal justice system. She was also providing food and education for forty kids with her own money. Within Bon Appétit, she was known as an activist who organized gala dinners for groups like Africa Outreach. So when my request arrived, Fedele knew who to contact. Karla's manager texted her to see if she could talk and she feared the worst. *Oh fuck,* she thought. *I'm going to get fired.* Instead her manager asked if she was interested in going to Puerto Rico. She said yes, of course, and ten minutes later he called back to tell her to get a flight and a satellite phone to travel out the next day. Karla arrived just as we moved into El Choli, and her experience in the catering world was essential to our operations there.

Karla was born in Mexico, moving to the U.S. when she was a teenager. Her father's family is from Santander in Spain, and her mother's side is a Mexican family that includes farmers. That mixture has given her a courage that I admire. She is humble but she also isn't fazed by people, rich or poor. In Puerto Rico, that's the kind of attitude that will take you a long way.

She learned her pastry skills in France as she dreamed of becoming a chef, at a time when her father wanted her to become a lawyer. At a young age, she had set up a pastry business in Mexico City, before going to culinary school. She did a *stage* with my friend Martín Berasategui, the brilliant three-starred Basque chef. She learned Italian cooking in Florence, before returning to Spain and another Basque restaurant in San Sebastián. Now she was working as an executive chef at DePauw University in Indiana for Bon Appétit. She liked to joke that she was the only Latina in Indiana, but I knew that was hard for her. People told her not to speak Spanish because it made them feel suspicious. Or they would ask her why a Mexican was telling Americans what to do, instead of washing dishes.

When you cook at scale, you become expert at processes, and Karla was a cooking systems specialist. She quickly helped establish order and structure out of the chaos of the first days at El Choli, where we had no power and little cooperation from the people on-site. She chased down suppliers and organized our teams to the point where we could ramp up quickly.

The day after she arrived, I wanted her to join me on a trip to Manatí, on the north coast, where we were going to meet some children and visit kitchens where we might expand our operations. I wanted to know if we were taking any food and water with us, but I was told we were just talking to the mayor and visiting some children's home.

"But what am I going to do if I find someone in the middle of the road who is thirsty or hungry?" I asked my team. "How am I going to help them?" We took a few hundred sandwiches and some water just in case.

They surely needed our help. As we arrived and were waiting to meet the mayor, I noticed two women dressed in their best clothes, as if they were heading to church. It was hot and humid already, at ten in the morning, and they were standing next to a water truck that had broken down. The women were politely but firmly asking the soldiers manning the truck why nobody had come to their community to bring food and water. Suddenly one of the women collapsed. As she recovered, she said she had barely drunk anything for the last two days. We gave her two bottles of water and some sandwiches to help her recover.

Further down the main road, we found a small restaurant, La Tacita, that had just opened with one dish on the menu: a mixture of rice, beans and a meat that looked like Spam in sauce. They had no power and were running a single lamp on batteries. They only had twenty-five portions to sell, at $5 a plate, because they had no money to buy more ingredients. I bought two plates and gave the owner more cash to help her get back on her feet, using the Estefans' cash. All she needed was some customers. In the meantime, cash was the smartest way to help. This was an island that wanted to feed itself, but the economy was paralyzed.

AS WE DROVE ACROSS THE ISLAND, OUR FOOD TRUCKS AND VOLUNTEERS were doing the same. By the end of our first week at El Choli, our food reached half of the island's seventy-eight municipalities in one day. It was a heroic effort, not least because we owned no trucks or cars ourselves, and bartered for our own gas.

After twelve days on the island, and after setting up three major centers of food production, the disaster was taking its toll on my body. I was exhausted and dehydrated. My voice was giving up, which made running a huge cooking operation hard. I was sleeping badly and barely eating or drinking in the sweltering heat. I lost twenty-five pounds, which maybe I needed to do. But it wasn't healthy, and my family and friends were getting worried.

I was tired and getting down. My pants were ripped, and my new Camper shoes looked like they were ten years old. I was stressed, but Nate was my rock, and Erin was learning quickly how to take over. My Puerto Rican team—led by Ginny and Ricardo—were on top of the details, and I had a whole team of chefs arriving from my restaurants in a few days, along with Compass Group chefs and chefs from our World Central Kitchen network of supporters. I felt we were strong, and I needed a break to rest and clear my head. The commercial flights leaving the island were all fully booked for weeks on end, so I emailed my friends back in Washington, and they generously helped me in many ways, including some who lent me the use of a private plane. I was so grateful for their support but felt guilty about the privileged life I was leading in comparison to the people on the island. I almost turned back at the airport, but Nate forced me onto the plane, driving away from the hangar so I had nowhere else to go. Herb Allen, the investor, was doing amazing work flying supplies into Puerto Rico and evacuating elderly people. I managed to get a seat on his plane to New York, and my friends Fred and Karen Schaufeld sent a helicopter to bring me to Washington.

I needed to regroup. But it still felt like I was leaving behind the Puerto Rican people I loved; the people who were still in desperate need. It was the middle of the night when I arrived home, where my wife and daughters were waiting up for me. I cried as soon as I saw them.

That night I met some old friends from Spain for dinner at Nobu in D.C. But I couldn't eat a bite. In my brain I was going through everything I had to do on the island. The next day, Fred sent a doctor to my home to treat me for exhaustion and dehydration.

CHAPTER 6

READY TO EAT

A MILITARY MEAL THAT IS "READY-TO-EAT" IS SOMETHING NO HUMAN being is ever ready to eat. Stuck on a battlefield, far from home or any kind of kitchen, an MRE (meal, ready to eat) may be a life-saver. But it is not a meal as anyone would understand it. The contents of a brown plastic MRE bag are so heavily processed and preserved that they only have a distant relationship with food.

Today's MRE is the latest industrialized solution to a problem the United States has tried to solve since the Revolutionary War: how to provide regular rations to the troops. In the Civil War, those rations were salt pork and a rock-solid bread known for good reason as "hardtack." For most of American combat history after that, the answer was canned food. But starting in the 1970s, the military began experimenting with something lighter. At first, it was freeze-dried food, developed by an Iraqi-American food scientist.[1] By the early 1990s, the experimental techniques evolved into something more like packed and flavored mush, in more than a dozen varieties, or what the military prefer to call "menus."

If you peel back the seal on its heavy plastic cover, you immediately realize that the MRE is designed like other weaponry.

It must survive impossible conditions like extremes of heat and cold. It must survive immersion in floodwater and the parched air of the desert. It must contain enough preservatives and packaging to last at least three years at eighty degrees, but even longer if the conditions are cooler. Why the need to last three years? Because the Pentagon needs to stockpile meals around the world in case of combat.

Inside you'll find some plastic-wrapped crackers and cookies. Maybe a plastic packet of nuts and raisins, and a powdered drink mix that needs water to make it liquid. The main course is a glutinous sludge that looks like it's been scraped off a sidewalk, while its ingredients sound like the contents of a laboratory. An official mark on the pouch must be some kind of joke: *Inspected for Wholesomeness by U.S. Department of Agriculture.*

This "wholesome" packet of calories (on average, 1,250 kilocalories) is skewed toward fat (36 percent) and carbohydrates (51 percent). Sadly for civilians and military personnel, it's also low on fiber, which means that people get constipated after eating the MREs for a few days. This may help on the battlefield, where there are good reasons to slow down the need to take a shit. But if you're not facing live bullets every day, you may not appreciate the constipation. For young people, this is hard to deal with, but imagine what it's like for the elderly. There's a reason why people joke that MRE stands for Meals Refusing to Exit or Massive Rectal Expulsion. Even for those on the battlefield, the MRE is only supposed to be eaten for twenty-one days.[2]

As the Pentagon puts it, the MRE has some well-defined purposes and requirements. "The Meal, Ready-To-Eat (MRE) is designed to sustain an individual in heavy activity such as military training or during actual military operations when normal food service facilities are not available," says the Defense Logistics Agency.[3]

How do you eat an MRE? Whichever way you want. You can

eat it cold, or boil the whole bag in water (if you have water and can heat it). If you can't do either, there's something called "a flameless ration heating device" packed into each bag that warms up the contents in ten minutes using water, which triggers chemical reactions.

They say that an army marches on its stomach, and that may be true. But a marching army also needs to carry its own food, and that helps define an MRE: its weight and dimensions are limited by the need to fit inside "military field clothing pockets." In reality, on a typical seventy-two-hour mission, packing nine MREs is too bulky, so the servicemen and -women find ways to rip out what they truly need and throw away the rest.

When all else fails, an MRE is better than going hungry. For civilians, including after Hurricane Katrina, it is certainly better than nothing. But it is no way to feed people for any extended period. I could see this firsthand in Haiti on that day when I saw children playing soccer with an MRE. If people living in desperate poverty cannot see these bags as food, who can?

In Puerto Rico, apart from the meals World Central Kitchen prepared, those plastic military bags of calories were the only food delivered in any reasonable quantities. MREs were the competition, and they were an expensive and soulless one at that. At current Pentagon prices, a box of 12 MREs costs $119, which is $9.91 per plastic bag of calories, not including the cost of shipping to the island and distribution across the island.

Our food was sourced locally to save money and help revive the local economy. Those boxes of MREs came from small towns and big cities in South Carolina, Florida and Ohio.

An MRE is a matter of survival. A freshly cooked plate of local food is a meal you're sharing with your family and community. I like to say that a hot meal is more than just food; it's a plate of hope. An MRE is almost hopeless. When you serve a plate of food, you gather intelligence about who needs feeding. When you dump a pile of MREs, you learn nothing about the true nature of the crisis.

Yet FEMA was determined to rely on MREs because it didn't want to rely on the Chefs For Puerto Rico, and no other nonprofit or contractor could produce the food. The alternative was handing out bags of chips and candy, or bags of uncooked rice and beans. At least the chips didn't need clean water and power before you could eat them. We were locked into a daily comparison between our chicken and rice, and a packet of chemicals that could be reconstituted to give the impression of chicken and rice. Only a government machine and the industrialized food economy could think that endless MREs were the way to feed millions of Americans for weeks on end. We had identified the enemy, and the enemy was an MRE.

OUR SOLUTION TO THE CHALLENGE OF CREATING A MEAL THAT WAS easy to transport and stayed edible for long periods was a simple, old-school idea: the ham and cheese sandwich. I have created many avant-garde dishes as a chef but there are few meals I'm prouder of than the hundreds of thousands of sandwiches we made in Puerto Rico.

Our sandwich line started out in the main dining room of José Enrique's restaurant, where we developed our methods, building on what we started in the Reef restaurant in Houston. Now at the arena, we could spread out into two huge sandwich lines. These lines were made up of young children and retirees, first responders after a long day's work and charity volunteers from the mainland, as well as many homeless hurricane victims who preferred to spend the day helping others rather than sitting inside a shelter. There were also off-duty members of the coast guard and officers from Homeland Security Investigations. They all became experts in making my ideal sandwich.

We sourced the white sliced bread in vast quantities from local bakeries, and bought equally huge quantities of ham and cheese slices from José Santiago and Sam's Club. But the key ingredient was the mayonnaise. Lots and lots and lots of mayonnaise, mixed with tomato ketchup for some extra flavor, and sometimes mustard too. At side stations, volunteers prepared huge bowls of mayo-ketchup mix, and cut the ham and cheese slices out of packets for quick assembly. Along two main lines of tables, other volunteers laid out slices of bread. Others set about dolloping mayo on each one, followed by a slice of cheese, a slice of ham and another generous slop of mayo. This masterpiece was finished off with a top slice of bread.

At the heart of this factory was my sergeant-major of sandwiches: a heroic volunteer called Dilka Benitez. Dilka was a wheelchair basketball player who helped organize the island's wheelchair ballers. Those organizational skills were clear in the sandwich factory, where she kept a close control of the volunteers, veering from encouragement to discipline in a few yells. She carefully managed the numbers and supplies, from the ham and cheese to the finished sandwiches. Dilka was helped greatly by two professionals: David Strong, my director of strategic initiatives at ThinkFoodGroup, and chef Griselle Vila, who runs her own catering company on the island. Together they built the volunteers into a hugely successful and productive team that changed every hour, as the volunteers switched in and out.

The sandwich line was one of my first stops when I returned to Puerto Rico after a few days of recovery back home in Washington.

I walked into the darkened chamber of the Coliseum where dozens of volunteers at rows of tables were dolloping mayonnaise and slapping ham and cheese onto endless slices of white bread. *Olé Olé Olé Olé, Olé Olé,* we sang together like the victorious fans inside a soccer stadium, as we just heard the news that we'd hit a huge target: twenty thousand delicious—and portable—ham and cheese sandwiches made in just one day.

"People of America! People of the world," I said to the volunteers and to all those watching what would become another video on social media. "You see the people of Puerto Rico feeding the people of Puerto Rico. Today, twenty thousand sandwiches. You see Puerto Rico together. The men of Puerto Rico and the women of Puerto Rico coming together. You should always be very proud of this moment. The first lady, she's my hero. She gave us the opportunity to use this space. I think the governor has been doing a great job. But more important, this is not about politics. This is about you, you, you, you, and you. One day, twenty years from now, you will be able to tell the story of how you together fed every man, woman and every child a happy plate of food. Thank you. Thank you. We love you. But I see we are not putting enough mayo."

As the cavernous room filled with cheers, laughter and applause, I turned to consult with my team: the paper bags we were using were sucking the moisture out of the sandwiches. We needed another solution to transport the sandwiches; one that wouldn't dry out the food as it sat in storage or traveled across the hot island all day. I suggested wrapping each one in a square of parchment paper.

"We don't want to just give food," I told my team as more boxes of sandwiches were carried to the door. "We want to give the best food. And please put more mayo."

I walked down the cinder-block hallway into the steaming kitchen, where a team of chefs were sweating next to giant vats of chicken and rice. My attitude was all about the protein. The trays of food needed more chicken, just like the paella pans outside. We couldn't short-change the hungry people of Puerto Rico. They needed the calories and they expected more chicken.

"We need to give people more food than we usually do," I explained. "I think it's important. I'm always questioning because people are hungry."

While I was back home, we had grown impressively. We created a pop-up kitchen in Guaynabo, west of San Juan, and served 5,000 people with chicken and rice. In a matter of days we went from 20,000 meals a day to 40,000 to now 60,000. We smashed through the milestone of 200,000 total meals, and were going to crush 300,000 on the day I returned. Our culinary school kitchens were going to open across the island in a matter of days, and I knew our daily meal numbers would spike again.

However, all our activity at the arena was now an unexpected and unwanted source of friction with its managers. We moved into El Choli with the support of the island's first lady, on a mission to feed the hungry. We thought, perhaps naively, that this publicly-funded space was a giant donation-in-kind. It wasn't. After a week of cooking, the management company SMG gave us an ultimatum, out of the blue: pay $10,000 a day to cover the costs of staffing and power, or leave in 24 hours. I was astonished. We were trapped, in a crisis, as we were trying to achieve the impossible. SMG wanted to back-date the bill to our first day there, saying they were a private company that did not get FEMA funding. The assumption seemed to be that we were awash with FEMA cash but the truth was that we were burning all our cash on food. We negotiated the cost down to $8,000 a day, but I was upset. My feelings about the arena were reinforced a few days later when the managers blocked us from expanding into more kitchen space. Domino's Pizza offered to open up its ovens to help our operation, using their own arena cooks. But SMG said they weren't allowed to, under the terms of their contract. The ovens stayed cold while the Domino's cooks joined our sandwich line. It made no sense to me, even though I know that arenas are complicated places with costs of their own.

The arena team was now boosted by more chefs, including several from my ThinkFoodGroup. Jennifer Herrera was a personal chef based in Dorado Beach who had grown up and trained in New York. Her mother was from Ponce, and she paid special attention to our satellite kitchen there. "It was a very personal experience," she said, "being able to cook *arroz con gandules, carne frita* and a substantial warm meal to warm their stomachs and warm their hearts."

Our hero was a twenty-five-year-old whose dedication and drive was an inspiration to us all. Alejandro Perez was executive chef at the Happy Crab restaurant in Dorado when he was diagnosed with Hodgkin's lymphoma. He finished his treatment a month before Maria and was warned not to go back to work. His doctors said any strain or accident could kill him immediately. His response: "It will be worth it. One life for 15,000 or 20,000 people I feed? It's worth it."

Initially Perez wanted to open a community kitchen in his home area of Bayamon, but the municipality said they couldn't support it. So he joined Chefs For Puerto Rico and came to El Choli to work long hours cooking thousands of meals. "My family thought it was risky and then they were concerned because I was doing crazy hours," he said. "Emotionally I felt useless to my family, to my wife. Especially my wife Carla. She left school because she had to pay for my treatment. It took a toll on me. But since we started here, she supported me all the way. She said she didn't care it has no pay. I just need to get out to do what I want. She even went with me to make sure I wasn't doing anything too crazy. She said, 'Go and do it. You are passionate about this.'

"I wish I could do this forever. I have never seen so many chefs, so many different people, work so well together, with one purpose, for one idea."

What inspired Alejandro to do all this? Meeting people on the island in worse situations. "I met this lady in Rincón whose house was gone," he recalled. "It was just a wooden frame. She was living behind a wall. She stayed in her bathroom through the hurricane as her house was getting ripped apart. But she was a community leader and she was helping everyone else. She was the one who needed help the most, but she was helping others. What we do here isn't one third of what has to be done. There are people sacrificing themselves. We're just giving our time and our skills. When I think things are going bad, I just go through my phone and look at photos like that. It's a reminder that it doesn't matter what goes on inside the kitchen. It's all for the greater good."

Erin organized a doctor to get him a checkup and the news was amazing: he was in full remission. When Alejandro returned to El Choli, everyone stood outside cheering, holding signs to celebrate his spirit and good health. It looked like another miracle.

EVERYWHERE MY TEAM TRAVELED, THERE WERE SIGNS OF DISTRESS, TWO and a half weeks after the hurricane. In Humacao, at the eastern edge of the island where Maria first made landfall, a simple sign by the road said it all: *La playa tiene hambre*. The beach is hungry.

While the island was still hungry, I drove over to the convention center for another mass care meeting. Our contract with FEMA was signed a day earlier, and would run for just two more days. But we had more than fulfilled our end of the deal. Since October 4, when the contract began, we had prepared and delivered more than 190,000 meals. By the time it was over, we would hit more than 300,000 on a contract that would only pay us for 140,000. The money wasn't the most pressing issue for us; the hunger bothered us much, much more. We didn't care about total numbers. We only cared about fulfilling all the orders we received.

It was my first time back at the FEMA headquarters since they kicked me out of the building at gunpoint. There was still an endless flow of volunteers from charities, newly arrived from the airport. Unlike the newcomers, who are issued credentials in no time, I still had no official ID card to enter the building, two weeks after the start of Chefs For Puerto Rico. Nobody ever told me why: it seemed petty and personal but that's how FEMA wanted to behave. So I was forced to grab people as they walked into the secure space—or rather, officials walked up to me to ask me what was going on outside their Caribbean green zone. Eventually FEMA official Elizabeth DiPaolo came down the escalator to talk to me outside the security line.

"I can feed 500,000 people tomorrow," I told her. "But I need to know what you think is the real need. We can use local kitchens and local food to get money into the local economy. I have already activated so many kitchens. I just need to understand how these contracts are going to work."

"The first contract with you is no problem," she said patiently. "That contract is already done. But we can't make another contract like that. That contract was just to get you started."

"But the requests we get are endless," I said.

"We know we need to do at least two million meals a day," she readily conceded. "But the people in charge are the state of Puerto Rico. We are all partners in support of them. In fact, we have to do six million meals a day. Work with us as a partner."

"Let me loose," I begged. "I can feed the island."

"The first contract was easy but if you want a second, it's something else. You've met your requirements already."

"It's the people of Puerto Rico who want food. And I can't provide it to them."

"So do you want another contract?" she asked, once again coming back to the bureaucratic needs, not the needs of the people.

"I don't need another contract. I already have people on the radio saying I'm getting rich from the people of Puerto Rico."

"Who is saying that?" Elizabeth asked, incredulous.

"The number one radio station on the island," I told her, knowing she had no idea what that was or where to find it. "You should hear what they say about FEMA and the governor too."

"Can you really do that many meals?" she asked, sounding just as incredulous as she did about the radio station.

"I have eleven kitchens already and can find more. There's a catering company at the airport and they can do more than 250,000. And I can do it cheaper and faster than anyone else. And on time."

The mass care meeting was about to begin, so we walked across the street to a huge windowless meeting room in the Sheraton hotel. Elizabeth opened the meeting by asking people to tell everyone their accomplishments for the day, hopefully with numbers, their plans for tomorrow and any challenges they might face to get there.

The Puerto Rico State Guard reported delivering 4.3 million bottles of water and 2 million meals, since the hurricane landed. The military was clearly the biggest operation on the island, and all of those meals were MREs.

The Department of Education reported they provided 115,000 meals to date through the first lady's "stop and go" distribution points across the island, including 2,050 meals that day. With the island's schools reopening tomorrow, there was a chance they could cook for many more. But the process of reopening and activating school kitchens was slow, and the education officials said it was taking time to get the orders through to the regional directors of the schools in the mountains.

Puerto Rican agriculture officials said they were still struggling to get enough truck drivers to move produce and water around. Their contractor had one hundred trucks but they were looking to activate many more. To me, it wasn't clear why the military couldn't or wouldn't help. Perhaps they were simply in the dark, but the Pentagon is never short of trucks or drivers. And the need was urgent: the next day the island's Department of the

Family was expecting to receive one million pounds of food from the U.S. Department of Agriculture.

At that point, the meeting turned to the nonprofits, or what they call volunteer agencies. FEMA mostly wanted to make sure we were all cooperating, or "playing in a big sandbox together," as they put it. But it wasn't clear what purpose or goals the group had, never mind how anyone could cooperate.

For now, the meeting was given over to the strange characters that disasters seem to attract. A new group had just flown in from Florida: Mutual Aid Disaster Relief, a fringe group with anarchist leanings that emerged after Katrina. Then the Scientologists said they had sixty pallets of goods coming from New York, and wondered how they could get them to Puerto Rico.

"FEMA doesn't ship donations," said one official. "We don't take the task of bringing donations over."

The Scientologists asked for help and the group pitched in with random ideas. FedEx perhaps? How about DHL? Maybe the airlines like JetBlue could help? Or they could get sponsored by a big corporation?

"We have the funds but we just don't have the actual transportation," the Scientologists replied.

I couldn't believe what I was hearing. It was not just totally detached from the crisis on the island. It was even detached from my experiences of solving problems on the island.

It was the turn of the American Red Cross to weigh in, as the organization whose unique charter gives it a congressional mandate to deliver disaster relief, coordinated by FEMA. The Red Cross report was a window into how inadequate the disaster relief was on the island. They said they had distributed 1.2 million pounds of food to date, which sounded like a lot until you divided it between 3.4 million Puerto Ricans needing 3 meals a day for the last 20 days. It was less than an ounce of daily food for each Puerto Rican.

Serving *sancocho* next to Chef José Enrique at his restaurant in Santurce, where the #ChefsForPuertoRico operation began.

Our first sandwich line inside José Enrique's stricken dining room.

A human chain of villagers in Utuado delivering hot meals from the Goya helicopter.

Wherever we served meals, I liked to talk to Puerto Ricans, especially children, about what life was like for them.

The Yummy Dumplings food truck was part of my Navy SEAL operations going to difficult areas like this one in Loíza—delivering food and gathering intelligence.

The cooks and volunteers at the Iglesia Jesucristo Monte Moriah, led by Eliomar Santana, standing above me.

Erin Schrode, COO of Chefs For Puerto Rico, was great at making real connections with the people we were feeding on the island.

I loved supporting the local food economy, especially when I saw fresh land crabs caught by my friend Papo, who did all the hard and dangerous work!

Delivering food to Don Lolo, a ninety-two-year-old veteran, who lives in Loíza on a street that remained flooded for weeks after the hurricane.

Whenever the food trucks stopped on their regular routes, there were long lines of hungry Puerto Ricans who came to rely on our mobile deliveries.

The destruction at Humacao airport.

Volunteers at one of the massive sandwich lines at El Choliseo.

Delivering asthma medicine across a collapsed bridge to Lilia Rivera in the remote town of Río Abajo.

With schools closed, or only open for a short time, children across the island were one of our top priorities.

Crossing the river to San Lorenzo was dangerous but vital in helping us understand the real needs on the island.

Our friends at Homeland Security Investigations were one of our earliest partners, distributing our sandwiches to some of the most remote corners of the island.

Even the measures of food were confusing and FEMA had no way of understanding what was going on. The Red Cross talked about pounds of food, while others were talking about pallets. We preferred to talk about meals, which was actually what FEMA's contracts specified. All these counts went into a big Excel spreadsheet that FEMA maintained and emailed every day. At the bottom of the spreadsheet, the total count of food was supposed to be there for everyone to see. Instead, the count was a calculating error because there was no standard unit of food that everyone used. If FEMA couldn't manage a spreadsheet, how could it manage an emergency?

"As part of our distribution efforts we are doing some feeding activity with prepared foods," the Red Cross explained. "Water continues to be a challenge and we're also distributing water filters, things like that. It depends on the solution of how we can get clean water to small communities and homes. We need to know what are the communities, and the needs and the long-term strategies because some areas will be without for a long time. We're trying to get creative right now. A lot of the normal tools in the toolbox are not there right now."

So there was no water pretty much everywhere, and they had no idea what the solution would be or when it would come. The Red Cross did say it had handed out some water filters for "almost 3,000" families. I couldn't understand why nobody was screaming about a water crisis for American citizens. Where was the sense of urgency?

The Red Cross was supplying a little food, they said. But the reality was, as they admitted, a bit more like people surviving on their own. "People are able to cook in a lot of communities," they reported. "There are a lot of grills and propane users. We're trying to support that effort as much as we can because we know there's delays in getting food distribution."

This is what we all knew from driving around the island, and what you had to assume in the absence of food riots. People were relying on their home supplies of food and propane. But who knew how long that would last? It was all very random and disorganized. And if you knew people were cooking like this, why would you continue to supply so many MREs?

The reality is that around 40 percent of Puerto Ricans qualify for what we used to call food stamps. On the island, this is called nutritional assistance but it is paid electronically on cards. Just 20 percent of the assistance can be redeemed as cash to pay for food, which is all of $80 for a family of four. With the lack of power and communications on the island, the supermarkets were finding it impossible to swipe cards, including the food assistance cards.

Why weren't the Red Cross and FEMA moving from the card system to good old-fashioned cash? Because the Trump administration was unwilling to rethink the basics of the system in this crisis. When the island's Department of the Family asked to increase the amount that could be drawn as cash from 20 to 50 percent, the Trump USDA said no. As a consolation, it said people could use the cards to buy hot food or sandwiches in approved stores. If you walked into a Puerto Rican supermarket, with empty shelves and refrigerators with no power, you would know how unrealistic that was. Even the Trump agriculture officials knew how bad the situation was. "We understand that at this point in time all food retail outlets in Puerto Rico are challenged by a lack of inventory, power and connectivity issues," the USDA wrote. "Additionally, ATMs are experiencing connectivity issues and limits on cash."[4]

As an exercise in mass care, the meeting was just that: an exercise. It was as disconnected from the reality of food shortages in Puerto Rico as the sushi bar downstairs in the gleaming white bar of the Sheraton hotel.

The Salvation Army, with all its resources, was typical. They said they had heard of areas of need, but didn't have the people available to assess those needs. In any case, their feeding numbers were in the single thousands. In Ponce, one of their largest sites, they were feeding a couple thousand people a day.

"Twenty thousand?" asked someone hopefully.

"No. Two to three thousand," the Salvation Army replied.

It was finally my turn to talk. "So we have our famous chef José Andrés," said Elizabeth. "Do you want to report on what you guys are doing today?"

I told them we were already distributing to sixteen municipalities, with four kitchens operational across the island, and preparations to expand quickly, including in Manatí in the north and Fajardo to the east. We were even feeding federal employees. "Today the National Guard called us in Toa Baja and they said they didn't have hot meals for the last two weeks, so we sent them four hundred meals of rice and chicken. Even though they kicked me out of this building," I told the gathering, which included military representatives. "The best partnership we have in the last two weeks, God bless them all, is HSI from Homeland Security. Many of them are border patrol and many of them are Puerto Rican. They have forty trucks, and they go to the rough and hard areas. They told me they wished they could bring food and water. These guys are very quick and they work on their own. Every day we've given them five thousand sandwiches to deliver. Every day. We have a fairly good track on where the need is because of them: which houses need help. So HSI has been great.

"If we can activate a helicopter operation, everyone wants to reach the middle of the island, but I already identified the people who can prepare twenty thousand hot meals precisely to deliver through army helicopters. It sounds so crazy but I only need to get to the army to say: Can you do that? The meals are prepared here by Puerto Ricans. People can receive 3,500-calorie meals anywhere that we want to land the helicopters. These are hot meals without having to open kitchens. At least this can be helpful for the next week or two until we are able to open roads in many areas. If anyone is interested in this, I have the intelligence on this and we could do 100,000 meals delivered by helicopter in twenty-four hours."

I told them about my maps, created by the army team, with all its details about resources. "This isn't just a dead map. This is a live map," I explained. "I'm dreaming we are coming to this map, put the numbers here, and everything will pop up here and we'll know immediately how many people we are feeding in every part of this island. So we can see the potential customers, people who are hungry, versus the potential of the givers, all of us. And we try to match the givers with the receivers. The army is doing a great job but I feel this could be a great map just to simplify our lives and look to see who we are, where we are, how we support each other and how we are feeding everybody in the areas that need it."

Another FEMA official talked about a different kind of exercise: last week he asked everyone to put sticky notes on a map to track what everybody was doing. Almost three weeks after the hurricane, the federal government was using sticky notes to manage information. The year was 2017, and everybody had access to Google Maps and Forms on their smartphones. But FEMA wanted people to stick paper on a map.

The main recipients of this information, FEMA made clear, was not the people feeding the hungry but their bosses, who wanted only the simplest of data. "We want to show it to leadership and say, 'These are the numbers you want to see.' They don't want to see a long report, they want to see a quick snapshot of what's going on," the FEMA official said. It wasn't clear what leadership he was talking about, but we all knew that President Trump didn't read his briefings and preferred pictures to words.

"The leadership wants to see quick and dirty," Elizabeth explained. "The need is 2.2 million meals on the island. Right now, we're doing 200,000. Our deficit is 1.8 million or—I can't do math. We are 1.8 million meals short. So that's why we need to get the urgency. This isn't going away. We're doing this much today and this much tomorrow. But this has got to be sustained over several months. So we really have to think longer term."

In fact, FEMA put the daily need at 6 million meals—3 meals for each of the 2.2 million people needing food relief. Of that 200,000 meals a day, my operation was already producing one-third. The other nonprofits were paralyzed or making only token efforts at providing food and water.

The remaining meals were almost entirely MREs from the military, but even those numbers fell far short of what was needed. According to the Pentagon, the military had delivered 7.7 million meals and 6.3 million liters of water since the hurricane, almost three weeks earlier.[5] Neither of those numbers was anything like enough for an island of 3.4 million people, most of whom had no clean water, no money to buy food, no supermarkets to shop in and no power to cook a meal.

"Chef, thank you," said one of the Scientologists, who started applauding. It was kind of him, but I found the meeting as depressing as anything else I'd seen in the disaster zone.

"You're the only one doing these numbers," Elizabeth told me afterward. "The only one out there."

That evening, Trump tweeted a video of federal efforts in Puerto Rico, saying, "Nobody could have done what I've done for #PuertoRico with so little appreciation. So much work!" The video began with a swipe at the media: "What the fake news media won't show you in Puerto Rico . . ." There were military helicopters lifting concrete slabs to shore up what looked like the Guajataca Dam, which was close to failing. There was the coast guard delivering medicine to the island of Culebra. But mostly the videos were a badly spliced together collection of unnamed agencies delivering bottles of water in unidentified places that looked like Puerto Rico. Perhaps they couldn't cite the numbers for food and water delivered because they didn't have a "quick and dirty" view, or because the spreadsheet counts were all calculating errors. The video ended, naturally, with Trump himself in Puerto Rico, including at his paper towel–throwing session, set to heroic music. Because the video was for one person, not for the American citizens of Puerto Rico.

I couldn't help but reply. "You've done nothing Sir," I tweeted. "The people at @fema @USNavy @USArmy @DHSgov @National-Guard are Americans in action! To lead you must be a follower . . ."

I returned to the AC Hotel, where my team of chefs from El Choli and our satellite operations was meeting at the end of a long day. My experience at FEMA made me feel we were out there on our own, in terms of our vision for feeding the hungry.

"We can feed the island," I told them. "We don't need anybody else. Everybody says we need the partners, but you don't need the partners. They should just leave it to the professionals. We can teach people how to do this themselves."

Karla had just come back from a trip to Manatí in the north and Mayagüez on the western coast, and her report was not good. We were exploring how to set up a satellite operation at the culinary school, but the supplies were not getting through.

"Oh my God," she said. "It's really, really bad. No communication at all. They can't even get products for their kitchens there. Can we at least do sandwiches for them? The only supplier that normally goes there is José Santiago, delivering on Wednesdays and Fridays. They only have a small walk-in freezer. But they have three grills, one oven and a lot of big pans we can use.

"We took sandwiches for the Salvation Army there. When we got to them, the Salvation Army said, 'Thank God. We were just going to Manatí. We don't have any food.' They need a lot of help.

We passed one house and I could see the closet from the street because half of the house was gone. We tried to get to one part of town, but the road was flooded. They didn't have hot food or anything. Maybe we could do paellas there? I was sobbing in the car. They are really struggling."

I explained to the team my solution for scaling up even more: an airline catering company I had met, called Sky Caterers. They could produce 300,000 meals a day with sandwiches, fresh fruit and chips. They could also make hot meals with chicken and fish, or vegetarian meals. Each meal would be packaged as it is on a plane. We could start delivering to the military whenever they wanted. And the meals would be produced in Puerto Rico by Puerto Ricans, helping the local economy along the way.

As for our sandwiches, I needed the team to maintain quality. "Increase the quantity of mayo," I said, yet again. "The paper bag is sucking them dry. That was a bad idea. We created the best sandwich in hunger relief, but this is drying out the sandwiches. Older people love the sandwiches but without a lot of water, it's hard for them to chew. They need to stay moist. Aluminum pans is how we began delivering them. It's so cool because they bring the pans back to you. If you put paper between the layers, they stay moist. I want people to say this is the first hunger relief operation where the food was good. Nobody at FEMA or in the NGOs talks about how good the food is. They only talk about how much food did we give away."

Our tensions with FEMA were worse than a difference of emphasis or values. Erin Schrode told the group how a FEMA official had come over the day before. "He wanted to check that all our volunteers were really our volunteers, because they weren't getting as many volunteers as they would like," she said. "They thought we might be getting them over here."

What he was really checking was how on earth we could

produce so many great sandwiches. After he asked about our volunteers, he stayed for an hour making sandwiches himself.

OUR CONTRACT WITH FEMA WAS OVER AND WE NOW FACED OUR BIGGEST test: we were producing huge quantities of meals for people who desperately needed them, but we had no way to pay for the ingredients it took to make more than seventy thousand meals in a single day. We had four kitchens open and were going to open another two, in Manatí and Mayagüez, tomorrow. We needed all the help we could get, and it didn't seem like we were going to get much help from either FEMA or the big charities.

It was time for me to contact my friends in Washington, so I started calling the senators I knew personally. I called Mark Warner, of Virginia, and told him that World Central Kitchen was a model. "We've been the spirit of how to give food relief," I said. "FEMA needs to cut the red tape. I've told them I'm pulling out unless they get me a contract. We are their best partner but they need to empower us to do more." The threat to pull out was a bluff: we were desperately short of cash but also desperate to feed the people of Puerto Rico. I knew we would never stop feeding them in their time of need.

I left messages for Kirsten Gillibrand, the New York senator, telling her FEMA had things back to front. You feed people first, and then you have meetings, I pointed out. Everybody is saying this isn't so simple, but it is so simple. If you are planning and planning, I continued, it would take six months before you fed anyone. What was my plan? Feeding the people of Puerto Rico.

Martin Luther King, Jr. used to talk about the fierce urgency of now to convince people to support the civil rights movement. Well, the urgency of now was in Puerto Rico. And the urgency of tomorrow is all over the world.

I was now spending far more time on funding and expanding our operations than I was on the meals themselves. We stopped

at the Federico Asenjo middle school in San Juan to meet with Julia
Keleher to check in on her order to activate the school kitchens
around the island. With more than 1,400 schools in Puerto Rico,
most of them with kitchens, the school cooks could easily produce 100,000 or even 200,000 meals a day. Keleher had essentially
told the schools not to worry: that feeding people was approved
and that food supplies would be coming. She even pushed the
National Guard to help with some food deliveries to her schools,
and the increased supplies seemed to be flowing. The department
said it delivered ingredients for an extra three million meals in
the weeks after the hurricane, but it was hard to see that on the
ground. I thought the schools could do much, much more if we all
worked together.

"This is the Department of Education, not the Department of
Food," she told me, in a humble way.

"You're actually more powerful than you think," I replied.
"Every school I visit has a great kitchen with great people and
great managers. That's what we want to see working for the whole
island. Ask them to double production, or whatever you can increase to. They know how to cook well. It's food from the heart."

It may be hard to activate more than 1,400 school kitchens,
but it's a whole lot easier than dealing with FEMA. My calls with
the agency were getting worse. I told them I could produce 100,000
meals a day; that they needed to partner with me. If they didn't,
I would go to the press. I would cook 100,000 meals a day and
then leave, saying I did it without FEMA. They told me I should
go raise my own funds. They complained they were hampered by
the Stafford Act, which governs FEMA, and said they just couldn't
cut the red tape. We would need to go through a bidding process
that would take several weeks. If people went hungry during the
bidding process, well, that was just too bad. We had no idea they
had already signed a $156 million contract for 30 million meals
that would never materialize.

"They are handing out shitty meals. Shitty, shitty meals," I told Erin. "They should be ashamed of themselves. And people are making money from this. They are making money from hungry people."

I decided to go to FEMA headquarters at the convention center for one last try. Without a security badge, I was stuck at the main door, looking at the many armed guards in the lobby. A state department official walked up to me and shook my hand. "I'm a big fan," he said. "Is there anything I can do to help? Anything at all?"

"Do you know, FEMA is bringing bread from Florida when we have twelve bakeries on this island? We are making fifteen thousand sandwiches a day when nobody else can produce so many," I told him. "And we're doing it with local bread, and delivering the sandwiches by local food trucks."

The official asked if I wanted to come inside, but I told him about the last time the armed guards had kicked me out.

Eventually I met with a couple of FEMA officials who insisted we needed to follow the contract bidding process. The conversation was depressing. Just a few days earlier, FEMA's regional director for the Caribbean, Alejandro De La Campa, had stopped by our Santurce operation to see what we were doing and to eat some of our food. He seemed impressed then, but I now realized it was just a façade.

As I walked away, I took a photo of the building and tweeted, "FEMA headquarters Puerto Rico, the most inefficient place on earth leaving the people of Puerto Rico hungry and thirsty." I tagged Donald Trump for good measure. I felt so liberated that I didn't have to go back there ever again.

I returned to El Choli and was reminded why we were there. It wasn't for FEMA and it wasn't about contracts.

Parked by the arena entrance was a small bus from the Cataño municipality, close to San Juan, carrying a group of volunteers. We had visited them a week earlier, and ever since, they had come to El

Choli every day to pick up one thousand meals for their community. They were my first big group of volunteers and their energy was infectious. As soon as I said hello to my Cataño friends, Lulu Puras from Mano a Mano walked up to me. She told me we needed to help the overlooked municipality of Patillas, in the southeast. "We delivered there last week and it was the first time they got food," she said. "When it rains, they put buckets outside to drink the water."

Governor Rosselló stopped by to see what we were doing, eleven days after he failed to show up in Santurce. I showed him our giant sandwich lines and walked him through the huge main kitchen. "We could do so much more with your help," I told him. But he barely said a word in response. As we crossed the arena floor, on his way out, I told him about the Rio Piedras neighborhood outside San Juan that needed seventy blue tarps. He agreed to give me the tarps on the spot, and I delivered them the next day. It made me wish that every problem on the island could be fixed as quickly and efficiently. It was also the last time I saw him or heard from him.

Whether it was the tweet or my threat to hold a press conference, FEMA emailed me to say they would renew the contract after all. "We have a really, really simple mission: to feed the people of Puerto Rico, and to feed them well," I replied, trying to hold out an olive branch. "That can't happen without the support of the federal government, which is responsible for the well-being of the American citizens here in Puerto Rico."

But the FEMA offer spoke volumes about their way of thinking about food relief and about Chefs For Puerto Rico. They proposed that we prepare just 20,000 meals for the next 21 days, for which they would pay us $8 a meal. In normal circumstances, you could feed people for half that cost, but this wasn't a normal situation: ingredients were expensive and we were cooking meals from scratch. By their own estimate, they were at least 1.8 million meals short every day. We were the only operation on the island preparing edible food, rather than shipping MREs, in any significant quantities. We were delivering to a wide range of places that were our partners, and returning day after day. MREs had no soul and just drove people away. While the overall value of the contract was large, at $3.4 million, the quantities of food suggested they weren't taking the crisis seriously. We prepared 73,600 meals that day, and were ready to open several new kitchens across the island. Their offer of twenty thousand meals was a sad joke.

I debated the options carefully with my team. We could use the money to produce more food if we wanted, not least because $8 a meal was more than we needed. But the principle annoyed me. They were over-paying and under-delivering, which seemed typical for a group of officials that had too much money and too little understanding. We decided to make a counteroffer: we would prepare 100,000 meals a day at $6 a meal for 11 days. The overall amount was $2.2 million more, but it was much more realistic, and would mean more food, more quickly, for less money. That was our final offer.

FEMA replied quickly to our final offer: they said no. We would go ahead with our press conference the next day. For me, this wasn't about the money. It was about some FEMA lawyer in Atlanta making decisions without assessing the very real problems on the island. I wanted them to commit to more as a reality check on the crisis in Puerto Rico.

"They used me for a week like *The Hunger Games*," I told my team. "The video they posted of me, explaining our plan, was just propaganda. That's how bad it is. They said they would take care of us and take care of the people of Puerto Rico."

I WOKE THE NEXT DAY FEELING FIRED UP BY WHAT WAS AHEAD. FEEDING Puerto Rico wasn't some act of generosity by the United States. The Puerto Rican people contributed their taxes to the U.S. government, and FEMA was one of the services they paid for. Now was the moment for FEMA to step up. I could deliver half a million meals tomorrow, cooked by local chefs for the local people. All FEMA had to do was empower us to do the job.

"Leadership isn't pretending to be a bad basketball player with a roll of kitchen paper," I told my team. "Leadership is saying that it's outrageous that people are drinking rainwater which is flooding the homes they sleep in at night. I can't believe I came to the island to feed some people and now we have to do a press conference to call out the federal government for the lack of response. I don't have any enjoyment in that. I am only doing this to push people.

"But there are people in FEMA who tell me, 'José, you are a small NGO. You have to follow protocol. We've already done more than you deserve.' More than I deserve? I don't deserve anything. The American people of Puerto Rico are in need of everything. I've been told by FEMA they have many priorities and food is not one of those priorities. How can they be serious, that food isn't the most important thing?"

With the help of my friends Bernardo Medina and Richard Wolffe, we called the press conference outside the entrance to El Choli, just in front of the giant paella pans. We stacked up boxes of fresh fruit and plates of freshly cooked food to show the world what we were preparing and delivering to the people of Puerto Rico. For good measure, we had on display an MRE in its brown plastic bag, as well as a typical meal from other NGOs: a bag of chips and an apple. I made sure my map was also on display, so people could see the full extent of our operations.

I wanted to show the people behind our operations too: our Chefs For Puerto Rico in person, as well as the local support we had built up. There was Henry Newman, a senator from San Juan, and Angel Perez, the mayor of Guaynabo. Lulu Puras from Mano a Mano joined me, along with my founding chefs Enrique Piñeiro, Wilo Benet, José Enrique and Manolo Martínez.

"So we're here today to talk about the food emergency that millions of Americans are suffering today here in Puerto Rico. You can talk all you like about other statistics in the island's recovery. But nothing is more important than food and water. Without food and water, there are no people to rescue and serve. Period," I told the reporters and cameras.

"The truth is that FEMA says it needs to provide more than 2 million meals a day to meet the needs of the people of Puerto Rico. That's just one meal for 2 million people going hungry. But by its own account, FEMA with all the partners and agencies including us, World Central Kitchen, are delivering slightly over 200,000 meals a day. And that's being generous.

"So at least 1.8 million Americans are going hungry still, every day, in the richest country on the face of the planet. And that's wrong. Three weeks after Hurricane Maria, and today is three weeks after Maria hit these islands."

I showed them the meals. "MREs are kind of the last resort for human beings but the first resort for the federal government. You cannot eat more than three of them without your stomach giving up.

"But then other people, because they cannot cook like we cook, they give this food and this equals one meal," I said, picking up a bag of cheese puffs and tipping them out on the table in front of me. "That's what we're trying to be feeding our fellow Puerto Ricans. And then we give them uncooked food like rice and pasta. But people don't have the money to buy the other things to go with that rice. They don't even have money to buy gas or gas isn't available. And if they do, they don't have clean water to cook that food."

I showed them the food that we cooked: the rice and chicken, beans and vegetables.

"We give them food that is done with love, that is done with heart, that is served hot. Made by Puerto Ricans, serving Puerto

Ricans," I said. We were on track to cook 100,000 meals a day by the end of the week, but we could be cooking 250,000 meals. "If we let the dogs out, we can feed the whole island," I declared, my voice now scratchy as I shouted above the constant hum of the generators.

Instead FEMA was happy with the way things were. They asked us to cook twenty thousand meals a day, and that was it.

"We still don't know who is going to feed this island," I explained. "Now, because of this press conference, FEMA says in a last minute effort, and after telling us it's over, we could do a few more weeks at 20,000 meals a day. Which is 100 times less than what this island needs. Or we can wait for several more weeks as people negotiate contracts. This is all about red tape, not feeding the people.

"Puerto Rico was hit by two disasters. The first disaster was natural. The second disaster is man-made by a clear lack of leadership. The sad truth is that FEMA is over-paying and under-delivering. It is paying too much for food, and too few meals are being delivered to the people. There is no urgency to the government's response to the humanitarian crisis. We only want to feed the people. Nothing more, nothing less. Because nothing is more important."

I thought the press conference was a success but my chefs weren't happy, for all the right reasons. They didn't understand why we stopped cooking for an hour, when we had so many people to feed. I walked into the arena kitchen and gave them a pep talk: we were all one team and we needed to show that to the world. We could cook even more, and feed even more people, if we had more support from the rest of the world.

THAT AFTERNOON WE WITNESSED FIRSTHAND WHAT WE WERE FIGHTING for. My friend Jorge Unanue at Goya Foods offered to take us into the mountains on his private helicopter, giving us a glimpse of the forgotten interior of the island—and what a food relief operation could look like if the military used its helicopters to help. For a few hours, the comfortable corporate Bell chopper smelled like a kitchen, with tray after tray of chicken and rice stacked on the floor. "It's like a flying restaurant right now," Jorge said. We dreamed about creating a Goya MRE with great protein, rice and vegetables, that we could deliver instead of the plastic military sludge.

It was a tricky flight to Utuado and we weren't taking any chances. We needed to stay clear of all the wires and poles that were down around landing sites. As we flew from San Juan into the mountainous interior, we could see the scale of the devastation. Vast numbers of trees were felled or stripped bare. Bridges were collapsed into narrow ravines and roads blocked with mudslides. High up in the hills, homes looked like they were on the brink of being swept away by the next mudslide. "If you want to feed people who are totally screwed, you have to come up here," Jorge said.

We found an overgrown baseball field on the edge of a hill where it looked safe to land. Beside the field was the remains of a gym: its metal roof peeled off and dumped on the basketball court like the twisted lid of a giant can of sardines. We circled a few times to see if there were any cables that might catch the helicopter, and set down slowly in the overgrown outfield. There seemed to be nobody in the town here, and the streets looked empty. "Don't worry," Jorge said. "They will come out once we land."

As we landed, my phone picked up a signal and I received some emails. One was from FEMA: they rejected our offer to prepare 250,000 meals a day. As I was watching people without roofs, people who hadn't eaten a hot meal in weeks, FEMA sent me an email declining my offer to feed all those people. The army wouldn't even give me helicopter rides to do the work, even though we heard they were idle, and I had to ask my friends to help me bring meals to places like Utuado. We could do this. But the federal government was refusing to let us feed the people. The *federal government*. The president and the director of FEMA. They should have been fired immediately for being so removed from the needs of the American people in Puerto Rico. They should have been ashamed of themselves. They should have resigned. I couldn't stop the feeling of helplessness and started to cry.

It was raining and the ground was muddy. Suddenly a few pickup trucks pulled up, and several men, women and children came to greet us.

"We have some food for you," I told them. "Did anyone deliver food to you already?" Just MREs, they said.

They made a human chain from the helicopter and helped us load up their pickup trucks with our meals, taking us the short journey to the middle of the town's main street.

We had landed in a sprawling village called Sabana Grande, close to the town of Utuado, where many people normally worked on coffee farms. But the coffee was destroyed by the hurricane and everyone was just focused on survival. There was no electricity and very little water, because the villagers relied on electrical pumps to get the water up to their homes. Three hundred families were living there, and they were pooling their resources.

"We cook one meal a day with propane gas," explained Norma Natal Rodriguez, a local schoolteacher. "We go to the supermarket to get rice, but there are no ATMs working and there's no cash. The banks are closed. They distributed water one day, but if you aren't home when they knock on your door, you don't get water. There's a lot of need here, but we can't go anywhere else. Especially the old people."

No power meant no water. No water meant no way to keep clean or flush a toilet. And yet, these people seemed patient, good-natured and generous. We started serving our meals from the back of a pickup truck, and everyone waited happily in line for a plate or two of hot chicken and vegetables, some mashed potatoes (which kept everything warm during the journey), some cold yogurt and fresh fruit. They ate it right there and made sure to spread the word to their neighbors who weren't at home.

It was hard to think of all the fear and guns in San Juan, at FEMA headquarters. All those terrified government officials had warned me about the dangers of traveling to a remote place like this. But the only unsettling thing about the families of Sabana Grande was why they weren't angry about their situation.

"You have no water and no electricity, right?" I asked Rodriguez.

"That's right," she said.

"And no one has come here to help you."

"Yes. But we are well. The community is united. Nobody bothers us," she assured me.

But how could she say they were doing well, with no food, no water, no power and no help? Because, she explained, the people farther up in the mountains had many more problems.

Her positivity was an inspiration to me in that moment. I was still reeling from FEMA's news. I had no words to express how I felt. And that was rare. America was able to overcome any challenge. It was able to beat the best of Europe and the world. We'd had bad moments in American history before and we were able to overcome them, to become better along the way. But this was one of those moments when we weren't better. The hurricane had landed twenty-one days earlier, but we still had no plan of how to feed our fellow citizens. A group of chefs had come together to feed 100,000 people a day and we could reach so many more. But now we were going to be shut down. Where were the senators and congressmen? Why weren't they asking more questions about how these Americans were living here?

"It means a lot that you have done this for us," said Eduardo Luis Piñera, one of the village's organizers.

"I'm here so that Washington knows there are Americans here in Puerto Rico that need help," I told him. "I know the governor is doing a lot, but the problem is that the central government has this bureaucracy that doesn't move."

Eduardo didn't seem to mind about that for the moment. "I'm sorry it doesn't look so good now," he told me. "But Puerto Rico is normally beautiful." I thought one of the most beautiful things about the island was its people.

We drove on to another part of the village, past a church whose roof had been ripped off, and beyond a dairy farm where the whole herd of cows had died in the hurricane. Telephone cables were down across the road, which was still drenched with rainwater. We flipped open the back of the pickup truck and started feeding whoever stopped by.

As the sun set, neighbors passed on word about our hot meals and a steady flow of families came by. We could see the community coming together before our eyes, smiling, chatting and eating.

We needed to head back before it was totally dark. Flying a helicopter was hard enough in these conditions without the added challenge of relying solely on instruments. We drove back to the overgrown baseball field and said our goodbyes. I gave them some solar lamps that you could inflate to create lanterns. Every time I handed them out, people were happier than when I gave them a plate of food. I never truly understood the power of shining a light until now. One old woman, wrapped tightly in a blanket despite the tropical humidity, insisted on hugging Erin tight. She treated us like we were her lifesavers. The night set in with no streetlamps or house lights to break the deep darkness of the island.

As we flew back to San Juan, the big city cast a distant glow but most of the homes and towns below us were as black as the night itself. The only signs of life were the white headlights and red taillights of the cars on the road. The rest of Puerto Rico was living in the dark.

CHAPTER 7

SEEING RED

SOME PEOPLE SPEND THEIR EARLY MORNINGS PLANNING HOW THEY ARE going to help the world. Others are just trying to get ready for another day at the office. Then there are the people who start their day watching cable TV news and tweeting in response.

That seemed to be the way Donald Trump woke up, on the day I was heading to a distant corner of Puerto Rico: the island of Vieques. As I was figuring out how to feed the people, with or without FEMA, the president of the United States was threatening to pull out of Puerto Rico altogether. First he blamed the islanders for "a financial disaster . . . largely of their own making," then he said their infrastructure was a disaster. Finally, he threatened to abandon the island, saying, "We cannot keep FEMA, the Military & the First Responders, who have been amazing (under the most difficult circumstances) in P.R. forever!"

It was not clear how or even why he would withdraw the U.S. military and first responders from American territory. Perhaps he still thought that Puerto Rico was a foreign country. Whatever he was thinking, it was heard loud and clear on the island: the Trump administration had barely made a dent in the recovery operations after three weeks, and now Trump himself was ready to quit.

I set out on my travels determined to make up for the lack of leadership in my hometown of Washington, D.C.

It's hard enough living on an island that has been devastated by two hurricanes and is struggling to get back to normal life three weeks later. But it's even harder if you live on an island off an island: a forgotten outpost of a limping colony. That was the fate of Vieques, east of Puerto Rico, where the full force of Maria first ripped through, before striking the big island. I had heard horror stories of lawlessness and despair there, in the initial days after the hurricane passed, but I wanted to see for myself what the situation was like. José Enrique had family in Vieques and Culebra, and we tried earlier to organize a boat trip, but the seas were too rough. It had taken some time, but we'd finally been able to rent two old single-propeller planes to take me and my team to the island, where we would set up our own World Central Kitchen food operation. "This is so important," said Karla. "The volunteers were tearing up about this trip."

Vieques is a beautiful Caribbean island with crystal-blue waters and soft sandy beaches. But it's better known for its colonial history than its tourism. The U.S. Navy seized two-thirds of the island in World War Two as a home for the British Navy in case the U.K. fell under Nazi control. For decades after the war, the navy used Vieques as a giant firing range and storage depot, even lending it out to allies for target practice. After years of protests, the navy withdrew in 2003 and the former bombing range became a wildlife refuge. There are fragments of a colonial history that stretches even farther back: the main town is called Isabel Segunda, after the Castilian queen who financed Columbus's trips to the New World in 1492.

I was still trying to fight my own political battles as we arrived at the small offices of MN Aviation at San Juan airport, where we could find our pilots and planes. The press conference was on the front page of the *Metro PR* newspaper, cast as chef versus FEMA. In a couple of days, House Speaker Paul Ryan was coming to visit Puerto Rico and I spoke to his office as we drove to the airport. I wanted to make sure they knew this wasn't a partisan dispute. We were the private sector trying to help fix a public problem. In many ways, we were the conservative solution to big government bureaucracy. There was a lot in our approach that Paul Ryan could support. His staffers listened carefully and promised to keep in mind my experience and advice as they moved forward.

We took two puddle-jumpers to Vieques so we could carry as much food as possible. Our pilot, his blue T-shirt and pants topped with a camo baseball hat, looked like he was going to play video games rather than fly between two hurricane-torn islands. We rumbled along the tarmac to a smaller runway, where we bumped and bounced our way to takeoff. Twenty seconds later, we were flying noisily over the deep blue ocean, bordered by tourist beaches that were entirely deserted. The luxury boats were all firmly docked in their harbors, and the golf courses abandoned. Inland, the rivers looked like they were full to the point of bursting. The land seemed saturated with water after two hurricanes and endless downpours since. Three weeks after Maria, the trees looked like early spring: a few small green shoots in place of the normally lush tropical leaves. After a few minutes of flying over the open sea, we turned sharply to the left to swoop down on a short airstrip along the coastline. Just two small planes were parked by the terminal, and there were few signs of activity.

My goal was to see how we could support as many as two kitchens for two or three weeks, until federal aid could take over. I heard the islanders here were getting MREs, but that was never a sustainable or desirable way to feed people. The challenge for us was how we would pay for this expansion. But I felt the urgency of now; the need to feed hungry people who were losing hope. We could figure out the funding later.

The airport terminal in Vieques did not reassure us about the state of the island. There was no power, meaning no lights or air-conditioning. The sole security guard was fanning herself with a tourist leaflet. Several windows were blown out and there were still sandbags outside the entrance. Gas deliveries to the island only came every three days or so, and when they arrived, there were lines that lasted three or four hours, as every islander tried to fill up his or her vehicle. Schools were only open in the morning, and the children returned home before noon, when the heat of the day became insufferable. The main employer on the island—the W Retreat & Spa—was closed for a whole year as they tried to clean up, repair walls and fences, and restore the landscape to something worth several hundred dollars a night. With four hundred employees, sustaining many more family members, the loss of the W was a body blow to the Vieques economy.

We drove past scores of downed trees and poles, and past piles of household debris heaped along the side of the road, to the town of Esperanza on the south coast. There we were delivering a truck-load of cooked food and uncooked ingredients to an empty beach restaurant, called Bili. Its chef, Eva Bolivar, was a friend of José Enrique and had just shown up at El Choli one day, knowing we wanted to help Vieques. She offered to assist us, and we prepared for regular deliveries of ingredients and cooked food, like today.

It was hot, and the task of unloading the truck was made much worse because the aluminum trays were poorly packed and stacked. They overflowed with sticky sauce, and the smell attracted flies as we sweated and heaved the trays into the restaurant. There were supposed to be dozens of volunteers to meet us, but communications with the island were nonexistent. We were on our own.

"We need to be smart," I told my team. "We're creating a mess."

We handed out some food to a few dozen passersby but the area seemed mostly empty of people. The few who were around were busy cleaning up the tourist businesses for some future date when the visitors might return. So we decided to load up a couple of pickup trucks and head to some homeless shelters and medical centers where we knew there were people in need.

The scene at the George Refugio was stark. Around a hundred people were camped on simple bunk beds, sleeping on plastic mattresses. It was clean and orderly, but lifeless: the only residents there in the middle of the day were the sick and elderly, watching television or taking their medication. There was a small kitchen that gratefully took our food, but they had good news: they were also getting food from the schools. It was the first sign that our work with the education department was succeeding. Two weeks earlier, we had first pushed Julia Keleher to order the school kitchens to help local communities. She was unsure if she had the power to do so, or if her schools had the capacity to cook for large numbers. She didn't even know if her orders were getting through to the schools because communications were so poor. But here on Vieques was the proof: a working model of how she could solve the problem of feeding those in the greatest need.

We drove back to the island's main hospital, where the hurricane had crushed any normal medical services. Outside, a giant Caterpillar generator coughed and belched black smoke as it struggled to provide only intermittent power to the hospital. Inside, there was no air-conditioning and the nurses fanned themselves with paper. The main building was deserted of patients, as the electricity cycled on and off in maddeningly short loops. Fans and lights turned on, then shut down. The air came alive, then it slowed to a swelter. The ceilings leaked water into the empty hallways. Outside, a wild horse wandered through the empty car park. The medical director, Dr. Betzaida Mackenzie, told me there was an old law that made it mandatory for the schools to produce food at a time like this. If that was true, I wondered why it hadn't happened from Day One. But food was only one of their many problems: they had set up a small air-conditioned tent outside the main door for a handful of emergencies.

We drove on, past more wild horses and the occasional stray cockerel, to the main town square in Isabel Segunda, where we set up our tables under a white canopy for some precious shade. The square attracted a sampling of post-hurricane Vieques. On one side were some watchful National Guard troops, standing outside the small government offices. Across the plaza on a concrete stage was a makeshift clothes rack with donations for needy islanders, and some tarps hanging to channel rainwater into buckets. One woman told us she hadn't showered since Maria, almost a month ago. Homeless veterans mingled with school-age kids, some with no shoes on their feet, as we started serving our food.

"We're not getting the quantity of food that we need. People are eating whatever they have left in their homes," Gypsy Cordova Garcia, the president of the city council, told me.

"They are trying to get food in the markets with whatever cash they have left. But it's only today that two ATMs were set up in the city hall. The bank was open but they couldn't update the information. They only had the information prior to the hurricane. So even if you have money, you couldn't take it out because they didn't update things. Government jobs are all that we have right now because the private industry is zero."

Many people had simply left Vieques, Garcia said, and he expected many more to follow them. "People are suffering," he said. "They are suffering from the lack of goods and services. But their spirit is good. We want to restore the island and we will get it done."

Sure enough, the line for the new ATMs, on the corner of the square, was as long as our line for food. It was also moving more slowly.

Next to our food table, my friend Roberto Cacho was demonstrating one easy solution to the island's problems: a Merlin Eco water filter, powered by solar panels. He put one tube into a bucket of rainwater, and it pumped out clean, drinkable water from another tube. The device could purify a gallon in just one minute. "They don't have a lot of things here, but they do have a lot of sunshine and a lot of rainwater," said Roberto, who was the original developer of the W Retreat & Spa in Vieques.

One of my best local chef partners joined us serving the food: Carlos Perez from El Blok. We left half the trays of food with him and told him to distribute them. I could trust him to get the food to the people in need.

Before we returned to the main island, I stopped by the local Boys and Girls Club to feed the children there. The kids lined up neatly and patiently to get a hot plate of chicken and rice.

"Who is hungry?" I asked. Everyone put up their hands.

It was such a simple thing. There was so little that needed to be done to improve people's lives. Looking at these hungry children, I couldn't understand why it was so hard to empower my team of chefs to make the people happy again.

My love of feeding the children meant we were late returning to the airport. We arrived to find that our plane had already left without us. We had to sit on the tarmac, waiting for another one to come back to get us. The place was almost deserted, save for a shack of a bar, tucked behind a couple of big noisy generators, where I bought a few cans of beer. We cracked them open and sat watching another puddle-jumper take a handful of passengers down the taxi lane to the end of the runway. Suddenly, as it passed by, the cargo door flipped open. We could see the bags inside, along with several packets of bottled water. I had to do something: the cargo could drop on someone, or unbalance the plane in the air. I started chasing down the plane, running down the tarmac as fast as I could in the tropical heat, waving my arms and shouting over the noise of the turboprop. Somehow, thankfully, the pilot saw me and he stopped to shut the door.

If I wasn't going to help them, who would?

IT WAS THE END OF MY THIRD WEEK ON THE ISLAND AND I WAS GETTING increasingly impatient with the big charities and the NGOs that had so much money but were doing so little. The Salvation Army liked to show photos of themselves handing out food, but they never said where the food came from. The Red Cross didn't care if the food was hot or cold. And then there was World Central Kitchen, with only three full-time employees across the world, now preparing more than 100,000 meals a day. We had crossed the half-million mark for total meals prepared, thanks to a small FEMA contract, a lot of small donations, and some credit lines and credit cards that were maxed out long ago.

The scene at El Choli was humming. The generators we all relied on were turning over and over to keep the lights on and the stoves hot. There was the constant clanking of trolleys carting ingredients in and trays of hot meals out. The paella pans were getting scraped with giant paddles to mix the herbs and vegetables with heaps of white rice. Empty vats were being washed clean with a hose by the food trucks. Volunteers used loud-hailers to shout to those waiting in line that their orders were ready for pickup. Giant fans tried in vain to cool down the cooks both outside the arena and inside the main kitchen. At the sandwich lines, more volunteers wearing blue gloves would cheer and sing when they broke new daily records. Along the walls were giant boxes of fresh mandarins: someone had delivered seventeen pallets of the delicious sweet fruit, but nobody knew why. A casual conversation with a supplier two weeks earlier had suddenly turned into a massive delivery of $70,000 of fruit that somebody had just signed for. Outside, our food trucks were waiting to be filled, alongside the Jeeps of our friends from the Homeland Security police.

FEMA was suddenly reaching out to me again, and it felt like the public pressure was finally breaking through. Two days earlier, the agency had sneaked out the news that its leadership in Puerto Rico was changing. The announcement got very little attention in the media, which had moved on to far more interesting stories. After the Las Vegas massacre, they quickly grew obsessed with Harvey Weinstein, the Hollywood mogul, and his disgusting sexual assaults. So maybe the reporters were too busy to read between the lines when FEMA said they had "expanded the leadership team overseeing Puerto Rico recovery efforts."[1] The regional director for the Caribbean, Alejandro De La Campa, was getting layered over by Michael Byrne, who would now be "federal coordinating officer." De La Campa would be working with local government officials instead. "While Byrne will oversee the current operational needs, De La Campa will focus on working with the mayors and their long-term recovery needs," FEMA stated.

It was a classic non-decision by FEMA: if De La Campa was good at his job, he should have stayed in charge. If he was no good, he should have been fired or moved out. Instead FEMA would now have two bosses: one in charge, and one just hanging around. No wonder the agency was failing to lead in this crisis.

By some strange coincidence, that change of leadership happened just two days before House Speaker Paul Ryan came to visit the island.

Along with Byrne came contact from Marty Bahamonde, a former head of external relations, who was one of the few FEMA officials in New Orleans after Katrina submerged the city a decade earlier.[2] Bahamonde was now director of disaster operations and he became one of my main FEMA contacts. On the morning

Ryan was flying in, Bahamonde emailed me to introduce himself and ask if I could see the new boss in an hour at the convention center.

We met outside, on the sidewalk, because of course I still had no credentials to get inside. Still, the meeting went surprisingly well. Byrne was positive about our work and asked me all the right questions. I wanted to know if he meant business, and it seemed like he did. He admitted that he was getting a lot of pressure from Ryan's office.

"José, you're doing an amazing job with no resources," he said.

I agreed that was true. We were even feeding the National Guard hot meals, as a small nonprofit. He put his hands on his head and shook it in disbelief.

"We can help you provide food to the island," I said. "I can take one problem off your shoulders. If you tell me and empower me, I can provide a quarter of a million meals a day for three weeks, taking the pressure off you and giving people the food they need. Then we can start moving away, if you feel the operation is under control. But right now, I've been in every part of the island and I can see this isn't under control."

He nodded and said, "José, there's a lot of bureaucratic things but I'm not here to tell you that you can't. I'm here to see how we can make it happen. Let me work on this. I'm good at that." Byrne was widely credited with clearing up the logistical logjams—especially at the ports—that were strangling the island's recovery.

We talked about my press conference, which had obviously poked FEMA hard. But I told Byrne I was careful in how I talked about FEMA. "I always talked about the great men and women of FEMA," I said. "I only said FEMA was broken."

"I thank you for that," he replied.

After our meeting I emailed Byrne immediately asking again for a new contract, this time for more meals over a longer time: 250,000 meals a day at $6 a meal for 3 weeks. It was much more food, at a much lower cost, than our previous contract or their previous offer.

He wasn't the only one suddenly trying to play nice with Puerto Rico. A day after threatening to pull out FEMA (along with the U.S. military, police, fire and medical responders) Donald Trump declared that he would never leave the islanders. "The wonderful people of Puerto Rico, with their unmatched spirit, know how bad things were before the H's. I will always be with them!" he wrote. It was a complete contradiction of his feelings of a day earlier, which made everyone wonder if he truly meant it.

I walked over to our sandwich operation and watched the volunteers laying out so many slices of bread, ham and cheese. They were well on track to break another record, making eighteen thousand sandwiches in a single day. We were producing more sandwiches in one day than the total meals prepared in our first three days in Santurce.

I started to cry. The truth was that we had no contract, and no way to cover our costs, despite all the conversations back and forth about numbers of meals and numbers of days. "We almost made it," I said. "We almost made it. We could feed this whole island."

I walked outside, into the bright sunshine and heat, toward the paella pans. I needed to lift my spirits and there was nothing like a few giant paella pans to do that. A song had been running through my head after hearing the paella cooks singing it a few days earlier. Some street musicians had serenaded them with a popular song that they had adapted with a few words to make it our own.

Voy subiendo, voy bajando: I go up, I go down.

Voy subiendo, voy bajando: I go up, I go down.

Tu vives como yo vivo, yo vivo cocinando: You live like me, I live by cooking.

Tu vives como yo vivo, yo vivo cocinando: You live like me, I live by cooking.

The original line was *yo vivo vacilando:* I live by slacking. But I preferred our adapted version, and so did my chefs, because all good missions need an anthem. We sang it loud and proud by the pans, and I felt my spirits lift. I started to believe that FEMA would come around.

I had been talking with a Washington friend, Jimmy Kemp, the son of Jack Kemp, who was Ryan's mentor and inspiration. Jimmy was a government relations consultant and he also ran the Jack Kemp Foundation, working to develop the next generation of leaders. He promised to help me navigate the Washington swamp of disaster relief, and I believed him.

Ryan's visit changed nothing and everything at the same time. He had just led the House to pass a $35 billion relief package a day earlier, including more funding for FEMA and some loan support for Puerto Rico. It was only a first wave of help, but it was still something to brag about. Ryan's public comments, alongside a small bipartisan group of visiting lawmakers from the House, were what you would normally expect from any visiting politician from the mainland. "We do not forget that these are Americans," he said. "A large number of them fight alongside us in our wars. I'll say it again: We are committed to helping Puerto Rico and the U.S. Virgin Islands get what they need to make it through this difficult time."[3] It sounded like a direct slap at the tweet from Donald Trump the day before.

FEMA was beginning to change; I could tell from the language they were using in public. At that morning's press briefing, they started talking about buying food from local suppliers and providing hot meals. It sounded just like our operation at World Central Kitchen. I was so happy.

I sat on some outdoor steps of the arena, in my oil- and ink-stained pants, smoking a cigar, trying to gather myself, when my phone rang. It was Elizabeth DiPaolo from FEMA. She told me the news stories from the press conference had put pressure on them all, and the agency wanted to figure out a sound plan.

"Can you really do this?" she asked, about the massive expansion I proposed.

"Do you remember that I have thirty restaurants?" I replied. "I know how to do this. I under-promise and I over-deliver."

It wasn't an idle piece of bragging. To feed the island, I needed to create the biggest restaurant in the world, and launch it in record time. My solution was not just to open more kitchens. At this point, we were operating out of ten kitchens spread across the island. To go from 100,000 meals a day to 250,000 meant adding a whole new factory and I had found the solution: airline food. The largest kitchen on the island was not at the arena, but at the main caterer for the airlines: Marivi Santana and her company Sky Caterers. Santana was introduced to me by Alexandre Vargas, a friend in Barcelona. Santana could produce in large quantities a boxed meal of a ham and cheese sandwich, plus fruit, water and snacks, for around $5. She could make hot meals at competitive prices too.

Santana's company had laid off its workers because the hurricane had disrupted so much air travel and tourism on the island. A partnership between us could feed the island and get her company back on its feet at the same time. It seemed to me like a great solution, and certainly much better than dumping more unwanted MREs onto the poor people of Puerto Rico. I didn't know if there would be a FEMA contract, but I was planning for a big expansion all the same. We were discussing the contents of the boxed meals and she brought samples to my hotel. Santana was conflicted, though. While she wanted to move ahead, her father—who was one of the owners—was resisting the deal for reasons that were unclear. She liked what we were doing and wanted to restart her business. But he had worked with FEMA before, and didn't seem comfortable that we had the inside track with them.

Still, we had sourced food and prepared it in huge quantities several weeks ago, when nobody knew where to look for food supplies, and when the island was in a far worse state. One way or another, we would meet the bigger needs. It was like opening a restaurant and not knowing how many customers would come in. Even if you start slow, you know you can build by doing a great job.

WE WERE RAPIDLY APPROACHING THE FOUR-WEEK MARK AFTER THE hurricane made landfall, and there were few signs that FEMA was adequately dealing with the humanitarian crisis. Above everything else, clean water—that most essential ingredient for life—was still in short supply. Shipping bottled water was both extremely expensive and difficult. The military had only transported 3.3 million gallons of potable water in the last month, which represented just one gallon for each Puerto Rican. In public, officials were boasting that 72 percent of the island had water. But access to water was different from water supply: without electricity to drive the pumps or filters, there was no water in the pipes in Puerto Rican homes. Only 15 out of 167 water treatment plants had regular power, and only 16 out of more than 2,000 pumping stations had power.[4] Those numbers squared with what I heard on the ground, as we delivered food. Water supplies—if they worked at all—would come and go. In any case, the official advice was to boil all water, if it emerged from the taps at all. That alone meant the water was undrinkable, since nobody had electricity to boil the water and most people were in very short supply of propane gas. On a tropical island, in the world's biggest economy, these daily struggles for survival were impossible to understand or accept.

My questions to FEMA were like a broken record: How many water tankers have you delivered? How many wells can you secure? Where are your water filters to clean those wells? Why don't you pay for water trucks to come from the mainland? The response was always a shrug of helplessness: we don't have enough cash, they told me. They would talk about contracts but I felt they were hiding behind these words. These weren't contracts we were discussing; they were lives. I began to think that we weren't even treating the people of Puerto Rico as well as we treat cattle. At least cows can eat grass and drink rainwater. But the island government was squabbling over whether the water was safe to drink or not, and the people were confused. On a Caribbean island, there are natural springs where all that rainwater emerges clean and safe for drinking. The official agencies needed to test the water and tell the islanders where to go. I couldn't understand why Governor Rosselló allowed this to happen. Why was he so reluctant to stop the squabbling so that his people could have clean water? What bigger priority could there be for any elected official on the island? Meanwhile, the federal government, with all its money, seemed not to care about stepping in and solving the critical problem of clean water.

I was crying a lot. Not because of exhaustion or the people I was meeting, but because of the inability of the federal government to help the people—people who have actually paid for that help. That is why Puerto Ricans pay taxes: to take care of things in the good times and the bad times. Here we were, a nonprofit offering private sector solutions to a Republican White House and Congress, and we were struggling to work with the government to help the American people.

There were signs that FEMA was slowly improving with its new leadership on the island. We heard back that they were now prepared to talk again about a second contract, which we desperately needed. We were preparing around 100,000 meals a day out of thirteen kitchens across Puerto Rico. Since the end of our first contract, a week ago, we had produced more than half a million meals. But we heard other news as well: FEMA was negotiating directly with our airline caterer for fifty thousand hot meals and sandwiches a day.

I met with my team in a windowless, cinder-block room, deep inside the arena. We needed to talk through this pivotal decision: Would we continue to grow, or begin to scale back? Could we continue to grow at all if FEMA took away our airline caterer? I didn't care who cooked the food as long as someone did: the priorities were feeding people, and reviving as much of the local food economy as we could. The challenge wasn't just about scaling up: it was delivering the meals to the places in need, with great partnerships and intelligence on the ground. We were burning through huge amounts of cash—as much as $400,000 a day—because of the number of meals we were cooking.

A lot depended on the true state of affairs with FEMA and the Red Cross, which was not easy to discover. We had asked for a contract for 250,000 meals a day for twenty-one days to meet the needs of the island. They came back to us with a fourteen-day contract for 120,000 meals a day. It was a big step up from their last offer of 20,000 meals a day, even if it wasn't close to what we could produce—or what the people of Puerto Rico needed.

"That's a feasible number," said Erin. "I don't know how we can really get to 250,000. We have no way to deliver that number, based on the actual infrastructure we have."

"That's why I've been talking to the airport catering woman," I explained.

Erin, who had been the point of contact with FEMA, said they would post the request for bids in the middle of the night. I didn't understand the secrecy, but we needed to be ready to respond. This wasn't a matter of money; it was a question of how we could continue to feed the people.

"The best thing that can happen to us is that somebody else picks it up, and our job is done," I said, only half joking. I had no idea somebody had done this already, and failed to deliver on the contract.

WITHOUT FEMA'S HELP, WE NEEDED SOME OTHER BIG SUPPORTER AND there was only one worth the trouble: the American Red Cross. I was trying to reach Gail McGovern, the CEO of the Red Cross, to see if they would help, but she wasn't responding. A week earlier, Gail had emailed me with what sounded like a complaint about me asking (politely) what they were doing in Puerto Rico.

"First, I'd like to thank you for all that you're doing in Puerto Rico to help people in need in the aftermath of Hurricane Maria as well as for your work in Texas after Hurricane Harvey," she wrote. "Disasters of this magnitude require the collaboration of many organizations, and the American Red Cross is grateful to work side by side with government agencies, other non-profit groups, faith-based organizations, businesses such as yours, and many other institutions to coordinate emergency relief efforts and get help to people in need."

It was interesting that she said the Red Cross was working side by side with other nonprofits, because I did not see that happening in Puerto Rico. The Red Cross wasn't just another nonprofit: it was effectively an arm of the government, based on its charter, its role in developing the mass care strategy and its shared position alongside FEMA in the mass care meetings. Yet, in all my travels across the island, I never saw a Red Cross truck or shelter or other operation. On top of that, her description of my nonprofit, World Central Kitchen, as "businesses such as yours" was just plain wrong.

"Mr. Andrés, I understand from a few different sources that you have expressed some concerns regarding the American Red Cross disaster response operation, most recently in Puerto Rico," she continued. "I would very much appreciate the opportunity to schedule time to meet with you—along with our head of Disaster

Operations, Brad Kieserman—to give you a fulsome picture of all that the American Red Cross is doing to respond to this spate of concurrent natural and man-made disasters, including Hurricanes Harvey, Irma, Maria and the horrific shootings in Las Vegas (and of course Hurricane Nate which is headed into the gulf this weekend). If you're willing, please let me know when you're available and I'll do my best to accommodate your schedule. On a personal note, it would be a privilege to meet you—my husband and I recently dined at the Mini Bar to celebrate his birthday, and it was an unforgettable evening."

I was happy that she enjoyed my restaurant. But I couldn't understand what all these other disasters had to do with Puerto Rico. Las Vegas was terrible, but that tragedy didn't involve a population of more than 3 million Americans without food, water and power for a month. The Red Cross had annual revenues of $3 billion and seemed helpless here, while my tiny nonprofit was feeding 100,000 people a day.

The Red Cross had said nothing after FEMA withdrew its support the week before. Today was the day I needed to talk to Brad Kieserman to find out if we could work together at all, ahead of signing any new deal with FEMA. We finally connected.

"First of all, you and your team are doing amazing work," he said. "My team has kept me up to date. I've paid attention to the work you're doing and it's amazing. I really do believe that what you are doing is incredibly generous of spirit. I would love to be part of your team and love for you to be part of ours.

"We haven't had a traditional food mission in Puerto Rico," he admitted, because of logistical problems of working on an island. Instead, the Red Cross was focusing on what he called "the distribution of emergency supplies," such as portable generators and hand-cranked chargers for cell phones. They were also driving satellite trucks around the island to help families connect. He said they had thirty trucks delivering these limited supplies.

"We're not able to deliver hot meals," he said. "We never have been. That isn't even in our mission. We don't have that equipment. But I will tell you that you are doing sandwiches that we could help deliver. We could be a partner for you. We could help."

That didn't sound like much help to me. We already had Homeland Security and hundreds of volunteers delivering our sandwiches, as well as our food trucks and satellite kitchens. Again I was surprised at how little the Red Cross was doing in Puerto Rico. And I was disappointed that there was no mention of them supporting us financially for the sandwiches they wanted to deliver. Still, I wanted to stay positive.

"Probably, Brad, this is a conversation we should have had twenty-one days ago," I said.

"Probably," he said, chuckling. "Probably."

"I've seen your people at work and I have learned a lot from your people," I said. "I saw them in Houston looking after five hundred people in a church. I saw your work delivering Southern Baptist church meals through your delivery trucks. Obviously I understand this is an island and you don't prepare food. The sandwiches, I think, are a great idea and it will be lovely that the Red Cross and World Central Kitchen bring them to the people of Puerto Rico. But the most important thing is the hot food operation. I wish we had here the Cambros that the Southern Baptists had. But I came up with menus that could be delivered one hour's distance and the caloric total is good to feed people.

"The question I always ask everybody is: From the federal government, the NGOs, the private sector, who is actually in charge of feeding the people? Because the people are hungry. There is real hunger out there. There are different reasons for that, to do with the economy, the lack of water and the lack of gas. But this is a conversation we need to have later, because we can't allow this to happen again on American soil.

"Maybe because of what I saw of the Red Cross after Sandy

and in Houston, I was expecting maybe the same here. You were always part of feeding people, even if you weren't cooking the food. But part of me thinks the Red Cross has let me down, and let the Puerto Rican people down. I have seen all the news about what the Red Cross is doing. But my perception of who has to feed Americans in an emergency always was that the Red Cross was the leader. Not necessarily in cooking the food, but in making it happen. I am having a hard time believing that now."

Brad sounded apologetic. "I understand that, I really do," he said. "And you raised a great issue. I'm connected to what is happening on the island. We ought to have a conversation on how we feed people in a disaster. But I will be the first to tell you that when people are on an island, this is a very different operating environment. Not because we want anyone to be treated differently, but because of the supply lines."

My experience had been different, of course. The supply lines were working, even if they weren't up to normal capacity. We were buying plenty of food. And the U.S. military could build supply lines to any part of the world in four weeks, never mind what it could do on its own territory.

"But I don't disagree with you," he continued. "Feeding Americans wherever they are in a disaster is a conversation—with your expertise and your network and your experience—it's a conversation you and I ought to have."

I pressed again for the Cambros, which would help us enormously in transporting the food. I must have been the largest consumer of aluminum trays in the whole Caribbean, but they weren't good enough. Sometimes they leaked, as they had in Vieques, and were not reusable. I had asked FEMA for some, and ordered more on my own, but they hadn't arrived. The least the Red Cross could do would be to ship over some of their famous red boxes. But Kieserman did not respond and seemed to care most about getting some sandwiches to deliver.

"I do believe there's synergy between the Red Cross and World Central Kitchen in the future," I said, "so we can feed people on Day One after a disaster, not twenty-eight days after a disaster."

We had prepared and delivered hundreds of thousands of meals, and now they wanted to deliver our sandwiches. I pushed him again for some financial support. "Is there a way for the Red Cross to help us financially?"

"I did check on the financial piece of this before I gave you a call," Kieserman said. "I'm being candid on this. We are not fund-raising particularly well off this. I have probably overspent my operating budget. I probably spent more money than we have raised, just to put satellite trucks on the ground. Like you, I don't get anything from FEMA. There was no free shipping in containers, and I used to work for FEMA. I was former chief counsel. We spent every donor dollar we got to purchase supplies. I hope that doesn't stop us starting this partnership."

I was shocked. I had no idea the Red Cross, with more than $3 billion of revenue, would only spend in Puerto Rico what it raised off the crisis in Puerto Rico. I thought they were the primary provider of humanitarian aid in any given American disaster. But their relief aid was defined by their fund-raising, not by the humanitarian concerns for the number of Americans struggling to survive.

"I was only checking my luck," I said. "Maybe if we went together as a unit two or three weeks ago, we could have been feeding the people of Puerto Rico faster."

I offered him some of our recent shipment of fruit and four thousand sandwiches.

"I can't thank you enough for the call," Kieserman said. "I know you said you are David and we are Goliath but honestly, where this operation is concerned, we are the David and you are the Goliath."

We ended the call and I threw my bottle of water at the wall.

McGovern quickly emailed to follow up. She assured me that Kieserman was fully authorized to make decisions for the Red Cross. But she also gave me no details of what a partnership might look like, and seemed much more concerned that I keep my mouth shut.

"I'm happy we were able to find a way to work together on behalf of the people of Puerto Rico," she wrote. "Now that we are partners, if you have any concerns, please bring them forward to us so we can work together internally to resolve them."

But we weren't partners and we had no deal or anything close to one.

"I don't think partners is the word yet, but I'm happy that we are finally collaborating on feeding the people of Puerto Rico," I replied. "In the weeks and months to come I will be open and frank of what I saw in Puerto Rico. It will be from a pragmatic side, trying to improve everyone's readiness and clear lines of responsibility. In my humble opinion specifically about the Red Cross, I always saw the Red Cross delivering hot meals to people in need in America. Even though you were not the producer but the delivery arm, it was an important role . . . So still no one has answered me this simple question. Who was in charge of feeding 2 million people?"

I assured her I was only speaking up for the sake of the American people who needed our help. "The federal government and the big NGO's, we let the people of Puerto Rico down and we need to make sure that the story never happens again," I wrote.

I didn't receive a reply.

There was a good reason why they didn't tell me more details about their operations. The claim that the Red Cross didn't have money for Maria was simply not true. According to the Red Cross itself, they raised $65.5 million for the victims of Maria. Of that money $18.1 million was budgeted for "food and relief items."

They said they paid for 8.5 million meals and snacks "served with partners," but that stretches the definition of meals to a point that most people could not understand. Even then, the Red Cross couldn't spend all the money it raised: its total spending came to $30 million, leaving $35.5 million unspent. The Red Cross generously gave themselves 9 percent of that leftover cash—or $3.2 millon—for their own general management costs.

During the six months after Maria, according to Kieserman, the Red Cross distributed 634,000 "bulk meals" such as rice or beans, fruit and vegetables. They distributed another 492,000 "prepared meals," including some of the sandwiches my volunteers made. The vast majority of what they distributed were what they called "shelf-stable" food boxes of MREs, candy, cereals, crackers and dried goods. The Red Cross claims the total number of those shelf-stable supplies came to 11.5 million "meals." Much of that food was provided by FEMA, the Salvation Army and others. In reality, the Red Cross was a distribution service running through a single food supply company, Caribbean Produce, which is a similar supplier to José Santiago. They had at one point fifty trucks on the road, delivering fruit and vegetables, as well as FEMA's stockpile of MREs and water bottles. We bought all of our fruit from Caribbean Produce but I never claimed their trucks—or José Santiago's trucks—were ours. I cannot understand this way of thinking.

Instead of feeding people, the Red Cross decided early on that its priorities were communications for families, some "portable power" for cell phones, water purification kits, and home repair supplies such as cleanup kits and roof tarps. After one month, they had two trucks that drove around the island allowing people to make satellite calls to their families and to charge their cell phones. It's not clear how charging your phone—either on a truck or with a hand-cranked charger—could make up for the lack of cell phone towers outside San Juan.

The water bottles were the heaviest items to transport, and even the Red Cross realized a month after the hurricane that the distribution of those bottles was not sustainable. "We did some calculations a few weeks in, and said if we keep distributing bottled water at this rate, and our partners continue to distribute bottled water at this rate, we'll have more bottles to dispose of than there is any room on the island to dispose of them," Kieserman said later, in what sounded like an echo of the environmental disaster in Haiti. "So we moved very quickly to buy water purification kits and water purification tablets."

The Red Cross spent $3.6 million on deploying and housing staff and volunteers in just one month, on top of $900,000 in salaries in three months.[5] At its peak, the Red Cross said they had five hundred volunteers on the island, having pre-positioned fifty before the hurricane made landfall. Many of those volunteers formed "assessment teams," gathering information from communities about what supplies they had or whether power was functioning. They even spent a colossal $3.4 million on freight and warehousing, a figure that includes the cost of moving supplies from their warehouses on the mainland—but not for the cost of the food delivered by Caribbean Produce.

It is curious that an organization like the Red Cross—normally so highly visible at times of a disaster—was so hidden from view. Its deliveries of millions of meals were invisible on the roads, and were not reported in FEMA meetings and emails, or relayed to the mayors we worked with. We had a $1.4 million FEMA contract while they had raised $65.5 million, which made them a very big David and made us a very small Goliath.

Did they ever consider supporting hot meals in Puerto Rico, as they do normally on the mainland with the Southern Baptists? "We considered our options every single day," Kieserman said later. "Each day, as we went forward, it was not my judgment that

the Red Cross attempting to replicate a hot meal model was going to meet the needs of the communities that we were trying to
serve, who weren't otherwise having their needs met."

Some nonprofits were delivering relief to Puerto Ricans the
right way. Mercy Corps (with annual revenues one-twelfth the
size of the Red Cross's) focused on the people in greatest need—
single-parents, mothers with young babies, the elderly and the
disabled—in the hardest-hit areas They delivered our food and
handed out thousands of debit cards with as much as $200 in
funds to help the islanders through. But they knew the cash wasn't
enough. "If we would give the cards alone, they would spend all
the money on water," said Javier Alvarez, their director of strategic response. "So we complement the card distribution with water
filters."

Mercy Corps normally responds to global emergencies, but this
season was different. "We responded to Harvey and this one because of the scale," Javier said. "We found people pretty down and
angry that the aid wasn't coming in as quickly as they wanted.
In most of the communities we visited, federal help wasn't there."
Key to their work was the intelligence they picked up about people
in need. For that, they worked with the University of Puerto Rico,
which had developed a vast reporting system called *Rescates,* or
Rescues, for volunteers to relay problems.

Compared to an international disaster, Puerto Rico was an
outlier for the worst reasons. Javier was struck by the contrast between the sheer number of people at FEMA headquarters at the
convention center, and the little amount of aid that was getting
distributed across the island. "In an international response, there's
a whole system of coordination and getting things out," he explained. "Usually we have a cluster system: we have a water cluster and a food cluster. It's very well organized. It's headed by the
UN, and humanitarian organizations like us coordinate and help.

Any emergency response is always messy. But it's a very strange situation where you see there are the resources, but they don't meet the needs of the people."

I WAS CONCERNED THAT FEMA WOULD NEGOTIATE THEIR POSSIBLE CON- tract directly with my airport caterer, Sky Caterers, rather than us. It's not easy to manage large-scale food relief: you need to be on top of the quality of the food, as we were every day, and you need to know how and where to make your deliveries. We had made many mistakes in the four weeks since Maria, but we'd also learned a lot about how to do this the right way in Puerto Rico.

"We're not here to get a medal," I told Erin. "We're here to push everybody to feed the island. If we lost the contract because the Red Cross or FEMA took our idea and signed with them, I would be glad for them to steal the idea. We're not proud. It's their role. It's their job. We're not here to feed people. Actually we're here to make sure people are fed."

Kimberly Grant, CEO of my ThinkFoodGroup, asked if the new contract wasn't a trap. Were they setting us up to fail by insisting we deliver water bottles that nobody could source? The military was tightly controlling water shipments as they arrived on the is- land. What if they audited us; did we have the paper trail to track every meal? And what if our intermediaries or partners were do- ing something unethical or unacceptable? These contracts were so big they could sink World Central Kitchen and put me at legal risk.

"I don't give a damn," I said. "They can trash my name and send me to jail. We're talking about liabilities when the people are going hungry and drinking water with animal pee in it."

"It's not just about us. It's the suppliers," she said. "It's the small guys. It's about them."

Kimberly is my CEO but she's a good friend too. She was pro- tecting me, my family and my company. And she had previous experience with federal contracts. I made the decision to feed the people. But I couldn't leave others to suffer the consequences of my unilateral decision.

"Let's make sure everyone is protected," I replied. "But they already promised they are going to be providing hot food. If we have to leave tomorrow, we'll tell everyone we ran out of money and ran out of support. We'll say it on Univision and Anderson Cooper and Fox News."

The next day marked a huge milestone: we were on the verge of hitting our one millionth meal, an impossible dream when we'd started cooking four weeks earlier, managing just a couple of thousand meals in a day. Now, we were arguing about the second million meals.

We weren't worried about audits. We had stacks of paper tracking every meal delivered, every request noted. We asked to see proof of identity from people picking up meals, along with a letter on official paper from the organization that wanted the food. From our first days, José Ortiz of Dame Un Bite had organized our paperwork, spreadsheets and forms. We tracked everything. But we were worried about how much money we were burning, producing so much food every day, and when we might get paid—if ever. On top of that, we were coming to the end of our short deal to operate out of the arena, as we neared the twenty-one-day mark of our operations there.

"We were never supposed to be here," I reminded my team. "We are here because the need is here. So we need to stop when we need to stop feeding the people of Puerto Rico."

If FEMA and the Red Cross didn't need us or want us, because they were feeding the island without us, that was fine with me. We could pack up and go home. But if they weren't doing that essential job, then we needed to stay. It was a simple thing, even in a world where we needed to embrace complexity to get things done.

There was talk of a third contract, possibly extending into

December, but it seemed very distant. If we were going to be here that long, cooking in such large quantities, we needed to look at expanding the satellite kitchens in a big way. These kitchens gave us tremendous reach, and I wanted them to serve food within a radius of one hour's travel. But not all of them could be expanded easily. And if we couldn't partner with the airline caterer, maybe we needed to think about buying a catering company ourselves.

I needed to know if FEMA was serious about the new contract, but I didn't want to ask directly. So I called Marty Bahamonde at FEMA and hung up. He called me back immediately.

"Sorry, Marty, I did that by mistake," I said, and he hung up.

"Actually, it wasn't a mistake," I told my puzzled team. "I did it to see if he returned my call. They're not going to walk out on us if we're busy and especially if we're humble."

Erin was worried that we could only scale up as fast as our suppliers, José Santiago.

"The food will keep on coming," I reassured her. "Santiago told me personally that they may have issues on this product or that product, but the food will keep on coming. We just have to be creative enough that if they don't have the mashed potato powder, we have grits. And if they don't have grits, we use wheat."

If we didn't get another contract, we would keep a skeleton operation going with small quantities in a few kitchens where there was the greatest need, like Vieques. And if FEMA negotiated behind our back with Sky Caterers, we could only hope that the caterer negotiated a good price, like we did. But in my experience, that wasn't likely. "They don't know how to negotiate," I told my team. "FEMA doesn't negotiate lower prices."

I called up my FEMA contact Elizabeth DiPaolo to see if she would tell me what was really going on. "Part of my plan to feed 250,000 people a day was bringing in this catering company at the airport, which I have never kept secret," I began. "I just heard from them and they said they just signed a contract with FEMA to start with 50,000 meals. I was very happy because it's happening and it's part of my plan of scaling up, using the resources of the island. Are you aware of that?"

"They are just doing cold boxed lunches," DiPaolo said. "I'm not aware of any hot food."

"They told me someone from Washington signed the contract with them yesterday for 50,000 hot meals," I replied. I knew they could prepare hot food and had trucks to deliver those meals. "The owner said he's working with FEMA because he says I've been a bad boy with FEMA. But it's not true. I've been a pushy boy. If I were FEMA, I would have hired them twenty-one days ago."

DiPaolo said she didn't know about that.

"It's like we are two different people negotiating to feed the American people," I said, before promising to come back to the FEMA offices to pick up a new version of my maps of the island. "I only want to be protected from getting kicked out," I joked. "I still don't have a FEMA card. It's a dream of mine. I will die disappointed without it."

CHAPTER 8

TRANSITIONS

WE PREPARED OUR ONE MILLIONTH MEAL, WITHOUT MISSING A BEAT, ON our twenty-second day of cooking in Puerto Rico. My original plan was to cook maybe ten thousand meals a day for five days, and then return home. But this was our third day preparing at least 100,000 meals in just a single day, and we had room to grow. We were cooking out of thirteen locations across the island and there was clearly demand for our hot meals and sandwiches.

We put out a press release and I shot a video at what had become my headquarters: the stretch of concrete outside the entrance to El Choli where the food orders came in and the meals came out. "Hello, people of America, people of the world. Today: big news," I began. "Twenty-one days in this beautiful island of Puerto Rico, and I can tell you at the World Central Kitchen Chefs For Puerto Rico initiative, we are about to reach today one million meals cooked by the men and women of Puerto Rico. Big day. I love you all."

Before Maria, today would have been the biggest day of the year for a totally different type of cooking. Michelin had just announced its star ratings for Washington restaurants, and my avant-garde restaurant Minibar had kept its two stars for the second year. I missed the call from Michelin because I was checking on the giant volumes of chicken and rice in the paella pans outside the arena. Whether the meals cost a few dollars or a few hundred dollars, you do your best with the ingredients you have. In the end, it's the same thing.

The milestone was a whole new reason for the media to remember Puerto Rico. The *Washington Post* declared in its headline that we had served more hot meals than the Red Cross.[1] A Red Cross spokesperson said they had "served" what they considered to be the equivalent of 1.6 million meals in the form of 150,000 MREs, 302,000 "meal boxes" and 1.4 million pounds of pantry goods like cans, rice and crackers. That might sound impressive until you remember that a single can of beans weighs one pound and retails for less than one dollar. The fact that they considered those meals, whether or not you could cook the rice, told you everything you needed to know about the leader in humanitarian aid in America. FEMA told the *Post* it had delivered 14 million meals, but that included our numbers, and the rest was mostly MREs. Those meals, by their own estimate, were only a little more than two days' worth of what the island needed in the three weeks since Maria.

We used the one million mark to ask people to donate directly to World Central Kitchen because the only limit on what we could achieve was cold, hard cash. We needed that new FEMA contract urgently. Our food costs alone were now running to several hundred thousand dollars a day.

The next day FEMA sent us an email notice to proceed on a deal that was good enough: 120,000 meals a day at $6 a meal for the next 14 days, worth around $10 million. It didn't meet the island's needs, at least according to FEMA's own estimates of 6 million meals a day. But it was substantially more food at a lower cost than our first contract, which had specified 20,000 meals at $10 a meal. It was an acknowledgment that our direction was the right one: pay less per meal to reach more people.

We met with our FEMA consultant Josh Gill at the penthouse bar of the AC Hotel, where he first found me a few weeks earlier. Gill was not happy, even though we were close to another contract, which meant that he was close to taking another cut of money that could otherwise be used to feed the island. He talked darkly about working with FEMA in Puerto Rico. He described a world full of political power, intrigue and ill feeling, and I couldn't tell if that was real or not. He was setting himself up as the only one who could navigate the swamp—at an offensively large fee—describing a strange organization that was both incompetent and powerful at the same time.

"One of my counterparts is extremely tied into FEMA and politically close to Brock Long," he stated, using words that were both so grand and so vague that I didn't really understand what they meant. "You have to be cognizant of the political capital out there, and it's a lot," he warned. "You need to be careful about the conversations because of word of mouth. The rumor mill here in Puerto Rico is the worst I have ever seen. And I have worked everywhere in disasters since Katrina, and this is the worst."

Gill also made it clear there was nothing we could do to recoup our costs for the lost week between our two FEMA contracts. They would not make the new contract work retroactively and we weren't willing to count our numbers inaccurately. The gap between the two represented more than half a million meals, or $3 million at FEMA's latest price. We would just have to swallow those costs ourselves. We were lucky that our work was getting such wide attention—through social media and traditional media—that we were attracting enough small and big donations to cover the shortfall.

For his part, Gill was mostly concerned with avoiding traps that could cancel the contract. He had visited a shelter and watched the delivery of our meals, which included donated pots of yogurt. "Are you sending utensils with the meal?" he asked. "I was watching people eat yogurt with their fingers. FEMA has in the past had a contract representative who goes out looking undercover. Little things like that matter. I just want to make sure someone doesn't go into a shelter and see people eating yogurt with their fingers."

FEMA was still insisting that we provide a bottle of water with each meal, but getting water bottles was difficult and expensive for everyone. "I saw a contract for 178 million bottles go out two weeks ago," Gill said.

I wanted to know if we could write a contract that said "up to 240,000 meals a day" instead of saying exactly 120,000. If we specified a high number, but took a few days to ramp up to that amount, we would be in breach of the contract unless the wording explicitly gave us the room to grow. Perhaps that could be the language of a third contract, if it materialized.

It was time to talk about Gill's fees. I needed to make it clear that his cap of $250,000 was for all his work, not for each contract. On a 14-day contract of 240,000 meals, he would otherwise take a huge fee of $1.68 million, even at the lower rate we negotiated of 50 cents a meal.

"It's not like every contract gives you a new one," I said. "I told you very clearly: you are going to cap at $250,000 for the entire amount. I think this is very fair."

"It's not fair," said Gill. "We have burned an exceptional amount of political capital."

"We can increase it maybe a little more. But we are just asking to be capped at 250. We negotiated before," I said.

"We also shook hands on the first one," he insisted, sounding increasingly aggrieved.

The first contract had already netted him $70,000. If we didn't cap him now, he could take home close to $1 million even with a smaller contract. Besides, there might be a third contract.

"You don't know there's going to be more contracts," Gill said. "You understand the political capital we burned. It's a lot. That's fine."

"You got me into this trouble," I teased him. "You know I was a good assist for you."

"You know we were a good assist for you," he shot back. "FEMA turned you down on the last email. They turned you down. I'm not going to argue because this wasn't the conversation that we had."

"For me, it's immoral," I said. "Too much is too much for everybody."

"It was 250 per contract," he insisted. "I need to discuss it with my team."

Gill walked off, leaving me with my team. We were arguing about Gill making an extra $70,000 from the two contracts, and my team feared that Gill would cause trouble at FEMA for me. Why take that risk for that relatively small amount of money? For me it was a matter of principle.

"It's a lot of money," I insisted. "It's money for doing nothing."

"For fifteen million dollars you can feed a lot of people," said Kimberly, my CEO.

"Is this the only guy who knows how to close the deal? It's like a bad movie," I said. "This meeting was supposed to happen in a dark alley."

WE DROVE TO THE HEART OF THE ISLAND FOUR WEEKS AFTER THE HURRICANE, and green shoots were just beginning to show on the otherwise bare trees. It felt like early spring in the mid-Atlantic; a seasonal change you never normally see on these green, tropical islands.

We were heading to Naguabo, on the other side of the El Yunque rain forest from San Juan, where we had opened our most unusual kitchen: in a church, not a culinary school or restaurant, staffed entirely by volunteers. One of the reasons for opening there was its isolated location, which meant that its meals could reach towns that would otherwise be hard to serve. That also meant a long drive for us, up and down single-track roads that scaled impossibly steep slopes.

After an hour of climbing and winding along narrow roads whose borders had collapsed into mudslides, we reached an octagonal church in the town of Peña Pobre, perched between several windswept peaks, overlooking jungle, lakes and small concrete homes. The eye of the hurricane had passed directly over these mountains.

On the lower level of the church, tucked under the main entrance, was a small kitchen where elderly volunteers stirred giant pots of rice, vegetables and meat over nine gas burners. The kitchen was lit by a single lightbulb from the ceiling, and one window to the outdoors. A serving hatch opened up to a long room with folding tables where other volunteers served cooked food from our aluminum trays. Supplies of ingredients—cans of peas and beans, bags of rice—were stashed in every corner and along every wall. The air was heavy with the sweet smell of broth, rice, chicken and vegetables.

At the center of it all was a young man who looked older than his years, wearing baggy jeans and a giant blue T-shirt. Eliomar Santana worked his days as a director in technical career education, but his real mission was as a pastor of this church: Iglesia Jesucristo Monte Moriah. Eliomar's smile was so bright you could easily miss his determination and compassion. Desperate to find help for his community of 270 souls, he'd heard on the radio about a chef cooking for people in need. He didn't know my name or anything about me, so he drove to FEMA's headquarters and asked there. They didn't know what he was talking about, but they told him about the chefs cooking at the arena, so he drove over there.

"I had in my heart that I wanted to cook," he said. "I tried to find someone to help us."

He first met our volunteers outside the arena, who said no. That wasn't our model: we gave people cooked food and only supplied professional cooks with ingredients to open kitchens. We had no cash to spare and couldn't just throw ingredients at places with no chefs, where we had no control over the quality of the cooking. Besides, we had a kitchen in Fajardo, which wasn't too far away, and they could help out.

Eliomar didn't take no for an answer. "Can I talk to your supervisor?" he asked.

Erin came outside and also told him no. "It's a little hard to do this," she explained, knowing I would say no too.

But then Eliomar talked about how much this meant to his community, and how they would all work on this together. "I want to do more than just give food," he said. "And my church will do the same. I want to cook."

Erin began crying and hugged him, saying she would make it happen. She didn't ask for my permission because she knew I wouldn't give it. Besides, I prided myself on creating a flat organization, where people could make decisions quickly. We had our way of doing things, and this wasn't it. On the other hand, this was a moment when people were really hungry. This was a time to let the dogs out, open more kitchens, and trust in the community. It was a good message.

Eliomar drove back to his church and found the lead pastor in the middle of a service.

"Let's stop the service and start cooking tomorrow," he told his boss.

The next day he showed up at El Choli with his entire church congregation: busloads of people who wanted to cook. That was the start of a food relief operation that brought his struggling community together. They began at 6:00 a.m. and cooked until 10:30 a.m., before serving meals every day at 11:00 a.m. They borrowed equipment from the community and they stored supplies anywhere they could find space, including in the bedroom of Eliomar's son. When they started asking for donations, his son broke into his piggy bank to hand over eight quarters to help.

"We take to homeless people, and we take to people who don't have food," Eliomar said. "We also take to other pastors to look after their community. People have no cash and no work. And the SNAP [food stamps] electronic system is broken. There's no money for gas and the nearest place to buy groceries is thirty minutes away."

Even the most basic services were gone. The area had no power and no water: the church team had to drive down to a local spring for fresh water and bring it back for the village.

We went upstairs to the sanctuary itself, where Eliomar had gathered his volunteers—around eighty people in all, many still wearing their aprons and mesh hats. They ranged from seven years to seventy years old. He introduced me to them and they started clapping, before he showed a video on a giant projector screen next to a sign saying *DIOS ES AMOR*: God is Love. I sat on the marble floor, my legs crossed, as the video showed endless photos of the church cooking and delivering food. It was too much for me to hold the feelings in. I tried rubbing my face and pulling my ears, but I started to sob. Heavily.

"In the name of everyone, thank you," he said. "It's a privilege you are giving your time to come here."

I thanked them all for letting us serve them. I told the story of how we grew from nothing, with the help of all the chefs and volunteers along the way. "It brings out the best in people," I said, "at this very, very hard time."

Eliomar then called his congregation to form a circle around me, holding hands together. "This is how we pray," he said. They all started saying their own prayers out loud, closing with a cascading echo of *gracias*. We ended up applauding each other and posing for a group photo. This trip was very moving for Erin and David Thomas, who both helped Eliomar to begin cooking. I gave them both a big hug. David and I had opened high-end restaurants across America. We were feeding the few in Beverly Hills and Las Vegas. But here we were, brothers in arms, feeding the many.

FEMA wanted to know how we could feed people and how we could deliver the food. This simple church showed how we could do it, and why we cared. It's better to give than to receive. But all FEMA wanted to do was receive. They were so, so stupid. We could have opened a hundred of these community kitchens, and there would have been nobody hungry. We could have produced enough food for everyone, and treated people with respect. This church was special but it wasn't unique. In Puerto Rico, you could find this kind of community spirit across the island as it recovered from catastrophe. People say that everybody in Puerto Rico wants to steal from you. That's not what I saw, and it's not true. *It's better to give than to receive.*

We walked outside, to the car park, where the volunteers were loading up aluminum trays of creamy soft macaroni and ham with sweet corn and peas, along with sausage and chicken. I rolled out my map on the driveway and marked the church as our newest kitchen. It was officially on the map. I pointed to a Salvation Army kitchen not so far away and asked the church team if they ever saw them or their food as the team delivered hundreds of meals in a twenty-mile radius from Peña Pobre. They said no.

THAT SAME DAY GOVERNOR ROSSELLÓ TRAVELED TO THE WHITE HOUSE to meet President Trump and his cabinet in the latest round of his quiet efforts to get more support from the mainland. In public, Rosselló was tactful and respectful, prodding Trump to do more while taking care to sound like he was praising him. When reporters asked the two leaders questions in the Oval Office, the difference between them was as wide as the island. Rosselló praised Trump excessively for the smallest things, while Trump praised himself excessively for things he imagined to be true. Neither man seemed connected to anything we were seeing in Puerto Rico itself.

"It's been a very, very difficult situation for many people, I will say that, and especially the island nature," said Trump, stating the obvious. "If you look at getting food there, we did. The distribution was very difficult because the roads were blocked and even the people of Puerto Rico couldn't get to their food, in many cases because of the distribution centers, and the roads were in really horrific shape, because of the storm, and sometimes because of before the storm. But with that being said, step by step it's taken care of."

So Trump knew people "couldn't get to their food" but thought it was because of the roads. And he now thought it was "taken care of." Either he was wildly ignorant of the food situation on the island or he was trying to fool the press and people watching at home.

A second question got to the heart of whether his information was accurate. Conservative media were jumping all over rumors that the cause of the food crisis was local government corruption: officials hoarding food for themselves. It was a nice way to give up, to pretend like there was nothing the government could do to solve this hunger crisis in America. Fox News reported that the FBI was investigating complaints that local officials were prioritizing supplies for their supporters.[2] The FBI said they didn't know if the accusations were accurate, but that didn't stop the story from going viral. I have no doubt there was plenty of corruption and favoritism, but they weren't the reason why people were going hungry in Puerto Rico.

In any case, Trump had made his mind up. "Well, I'm working very closely with the government on that because there has been corruption on the island and we can't have that. You know, we're sending a lot of supplies, we're sending tremendous amounts of food and water and everything." Washington was sending tremendous amounts of MREs, but both the food and water were totally inadequate. Rosselló said they were investigating "whether there has been mismanagement of food" and promised "there is going to be some hell to pay" if it turned out to be true.

As for his own federal efforts, Trump seemed to think they were delivering food and water by helicopters in great quantities. "You have areas in Puerto Rico where we literally had, and still have to—but it's getting less and less—deliver food and supplies by helicopter because the roads have been wiped out and the bridges have been wiped out," he explained. Perhaps he watched his own White House video too quickly to understand what the helicopters were doing and how many of the shots were repeated. Because I, for one, had had no luck getting military helicopters to deliver food. And I heard from many people in the military that the helicopters didn't have enough missions to keep them busy.

One reporter asked Trump how he rated the White House response on a scale of one to ten. "I'd say it was a ten," he said modestly, pointing to comments from James Lee Witt, who ran FEMA under President Clinton. Witt said he'd give them an A+ for their work on hurricane recovery. "I think we did a fantastic job and we are being given credit," Trump said. "It was very nice that the gentleman who worked for Bill Clinton when he was president gave us an A+, and that included Puerto Rico. Gave us an A+ and I thought that was really very nice. And I think—I really believe—he's correct. We have done a really great job." Witt later issued a statement making it clear that he was only talking about the response to the earlier hurricanes in Texas and Florida, not the response to Hurricane Maria in Puerto Rico.

Rosselló was asked the same question, and he dodged it by promising that power would be restored to half the island by next month. Trump wasn't satisfied and pushed the governor to praise him. "Did the United States—did our government—when we came in, did we do a great job?" he asked. "Military, first responders, FEMA—did we do a great job?"

"You responded immediately, sir," Rosselló said, dodging the question once again. "But if you consider that . . . we've gotten about 15,000 DoD personnel in Puerto Rico, about 2,000 FEMA personnel, HHS and others—the response is there. Do we need to do a lot more? Of course we do. And I think everybody over here recognizes there's a lot of work to be done in Puerto Rico."

Rosselló was too subtle. Trump ended by going back to the comments of the former Clinton official for his A+ rating. "While I don't know him, I would like to thank him for what he said."

I wouldn't give Trump or Rosselló a ten or an A+ for their work. And I certainly wouldn't give myself anything more than a five for my own, because there was so much more we needed to do. We failed to reach so many people who needed so much help. The only people who deserved a ten were the volunteers and first responders who were so selfless in their work. But as leaders, we did not deserve anything close to a top grade.

I LIKED TO LEAVE SAN JUAN AS MUCH AS I COULD. NOT JUST TO ESCAPE the intensity of the giant feeding factory there, but to pick up intelligence on what was happening around the island, and how our satellite operations were doing. I needed to see for myself where the real needs were, and how we were meeting them. Over time, people were surely adjusting to life after the hurricane. There would come a time, probably soon, when we would need to change what we were doing because life on the island was changing too.

This was one of the most important reasons we were so different from what I saw of FEMA and the other nonprofits. Our intelligence operations helped shape what we were doing and where we were doing it. Their intelligence seemed at best out of date and at worst nonexistent.

I liked to visit our culinary school kitchens to check on the quality of their cooking. With eight of these running, they were producing large numbers of hot meals and needed regular visits to make sure they were getting the right amount of chicken on each plate. It was too easy for them to skimp on the expensive ingredients to save themselves money while denying the people of Puerto Rico the good meals they deserved.

But my best trips were with what I called our Navy SEAL operations: our food trucks. They weren't perfect. They were old trucks and prone to breaking down, so the distances we asked them to drive every day were a heavy strain. They drove into areas where the Homeland Security police only traveled with guns and flak jackets. But our food truck partners were determined to serve the people in need, and they never experienced any trouble as they handed out up to 1,500 meals a day. By now, a full calendar month after Maria, they knew their routes and their communities very well. They were friends with their customers; they knew their family stories and their daily schedules. They had learned what times of day were best, and where the elderly were housebound and couldn't come to the truck, and which kids could help them serve the food.

We had ten trucks in operation but two of them were the heart and soul of our operation, run by two sisters, Xoimar and Yareli Manning. They were two of our original partners from day one of our operation at José Enrique's restaurant in Santurce. José Enrique's text to join us was one of the first they received after several days of getting no cell phone signal following the hurricane. It came just in time. "I was going to the States and moving out of Puerto Rico," Xoimar said, "because I couldn't handle it. I wanted to be able to run my business. My generator wasn't big enough to run my kitchen. I had no power, no water, no signal, no school for my daughter. My friend in Florida told me to come to her house. I'm frickin' dead and this is the end of the world. I thought I was leaving as soon as I could."

The sisters started their food truck businesses a decade ago, after watching a TV show about a food truck in North Carolina called Chirba Chirba dumplings. Xoimar's husband turned to her and said, "This is what we're going to do." The food truck scene was just starting up, and they bought a truck the next day. The business, Yummy Dumplings, was a huge success, running six days a week with double shifts on Thursdays and Fridays, until the day that Maria struck. Xoimar ran Yummy Dumplings and Yareli ran The Meatball Company, normally out of a food truck park they started in San Juan. Yareli had visited all my restaurants in Washington, D.C., and never thought we'd end up working together.

They were fearless and tireless, driving for hours each day in old trucks with no cell phone service, into places where our Homeland Security friends expected to see lawlessness. Instead they found families who were struggling to survive and delighted to see them. "This gave me hope that things would be getting better. We were working and we were busy. It was better than sitting at home. It felt so good and so right. I'm willing to do it forever," said Xoimar. I loved their spirit and no-nonsense attitude, and I especially loved their family. Xoimar's ten-year-old daughter, Lola, reminded me of my own girls, not so long ago, and she showed up to work every day without complaint. I would tell her to take a break from the sandwich line, but she would never listen. I was so impressed with her, I promised I would pay for her college education.

The rundown streets of Loíza, on the northeast coast, were

the regular food run for the Yummy Dumplings truck. "This is my dad's neighborhood," Xoimar told me, as we walked the streets making deliveries behind the truck. "This is where he spent his time when he was young. So for me, serving the people he grew up with is important." The local kids ran out to greet her and the older folks embraced her at every stop. They had only seen their local mayor once since the storm, handing out some diapers and water. Yummy Dumplings was the biggest provider of storm relief in the area. The food truck team parked near a baseball field, and whistled to announce their arrival. Dozens of kids, some on bikes, came from all points for a plate of mac and cheese, rice with beef stew, an apple and a bottle of water. "I come twice a week and I try to give to them a different meal each time," Xoimar said. "These people are having a hard time."

We drove on another few blocks to where some people were still stranded by floods. One of them was ninety-one-year-old Teodoro Figueroa Rodriguez, known as Lolo, whose entire front yard was under water. I put on some rubber boots and waded through the murky water to take plenty of food to his marooned home. Lolo was a National Guard veteran who needed the help. His daughter worked long hours as a nurse in San Juan, so he was alone for most of his day.

"Thank you for looking after me and not forgetting me," he told me. "I'm glad the young people like you care for people like me."

The next day I wanted to find an even harder-hit part of the island to see how it was faring. For that, my Homeland Security friends strongly suggested they drive me around, just in case we found some of the long-rumored but never-seen dangers they were so heavily armed for. I was happy to say yes, because we needed another car, so I climbed into a Jeep with deportation officer Alex Sabel and special agent Krystle Intoe. With plenty of sandwiches and bottles of water in the back of the Jeep, we headed for Aguadilla, two hours to the far west of San Juan, where conditions were supposed to be very poor. But halfway there, we decided to stop at another poor town, Morovis, at the center of the island, just to try our luck as we searched for pockets of hungry people. In the middle of the town we spotted an SUV parked by the side of the road with a couple of local officials handing out MREs. Nobody was interested in taking them. We offered them our sandwiches to distribute instead, and they happily took them.

A little farther down the road, there was a long line of people outside a simple tin shack with smoke billowing out from under its corrugated roof. My gut told me this was exactly what I was looking for: a sign of food life returning to the island. The smoke was coming from a charcoal pit where ten giant spits were slowly turning, driven by a long bike chain, as they roasted dozens and dozens of chickens. The smell of the chicken and adobo spices, dripping onto the charcoal, was enough to make your mouth water from a block away. A whole chicken with a plate of beans and rice cost less than $9.25 and could easily feed four people. That was several dollars less than each of the single-person MREs they couldn't give away nearby. This was food that the people of Puerto Rico wanted to eat, at a price they could afford: just a couple of dollars for a delicious and filling meal. In fact, it was so good, it was food anyone would eat in Washington or New York at prices two or three times higher.

The restaurant, called Maelo's Chicken Fever, had only just reopened and it seemed to be doing very well. I asked the customers lining up if they knew where people were struggling nearby and several pointed me to the village of San Lorenzo. The bridge there had been washed away by the floods that followed Maria, and they were effectively cut off from the rest of Morovis. The old river crossing was a ford and if the water flow wasn't too strong, it was possible to drive across.

"This is good news. Things are coming back to the old Puerto

Rico way of life," I told my HSI friends as we ate a couple of plates of delicious chicken. "But there are communities that are isolated and we need to have the urgency of now to keep bringing food and water. The food issue is still with us."

Our mission was to find these people and to help them. Aguadilla was just another one of our satellite kitchens, where I would be checking the quality of the cooking. This trip to San Lorenzo seemed much more important. We drove another ten minutes to the river, where the huge concrete bridge was collapsed. Part of it was beneath us on one side of the river. Part of it was washed away on the other riverbank. A middle section was nowhere to be seen.

Down below, the ford was clearly visible but it ran alongside a steep drop in the riverbed, and the currents seemed too strong and turbulent for our Jeep. A couple of people were crossing by foot along the edge of the ford, but as they neared the middle of the river, they all stumbled to knee depth as the ground grew uneven. It looked like a tricky walk at that point, with the drop of the river so close, the water flow so strong and the riverbed so thick with slippery moss. One false move and you'd go tumbling headfirst into the water. Overhead there was a wire strung from one bank of the river to the other, but it didn't look like it would offer much support. It was a long way from where the people were walking, over a part of the river where the flow looked even faster.

We waved to some of the people crossing the river by foot and they told us around a thousand villagers lived in San Lorenzo. We met them in the middle of the river, gave them the few boxes of sandwiches we had left, and they walked them back to the other side. I decided to return with cooked food for all the village. For now, we could either drive over to our kitchen in Manatí, where they weren't expecting us, or we could head back to the roast chicken shack and support the local food economy. It wasn't a tough choice.

Alex, my Homeland Security escort, helped me carry the sandwiches halfway across the river to the people walking to San Lorenzo. "I'm forty-six years old and this is the craziest thing I've ever done in my life," he said.

Alex was a deportation officer but an unusual one. Born in Argentina and raised in Colombia, he now lived in Phoenix, Arizona, where he worked in peer support for people going through trauma. "We have these law enforcement skills to help the average person. You just have to open the window a little bit more," he explained. "A lot of us are immigrants too, you know."

"I never thought I would be helping people by working with ICE," I joked. In fact, I had staff of mine back on the mainland asking me why I, an advocate for immigrants and immigration reform, was working so closely with the people behind all of Trump's deportations. But photos on Twitter only tell half the story. I could only say that even though ICE was breaking up families, there are some good people there who are doing good work.

As we drove back to Maelo's, I regretted not fully crossing the river. I wanted to see San Lorenzo for myself and talk to the people about how they were doing. Then by chance I saw my opportunity: ahead of us on the road, by some backhoes clearing trees felled by the hurricane, was a convoy of three military Humvees.

I jumped out of our Jeep to talk to the group inside them, a small and unlikely combination of military police and air force engineers. They were stationed at the same base and decided one day, out of boredom, to see what they could do together. The engineers were specialists in water systems and the MPs knew there were water treatment plants that needed fixing. Now they were just about to repair the local water treatment plant, to bring it back online. I asked if they could help us cross the river in their Humvees, and they agreed, once they completed their repairs in a few hours.

"If we only let people in the field do what they need to do, like you and I and these men and women are doing, things will be operational so much easier and so much quicker," I told them. "If we had to coordinate with headquarters, it would take three days." Or it would never happen at all, I thought.

We took their satellite phone number and returned to Maelo's to get our chicken. A lot of chicken: five hundred portions, in fact. The restaurant was delighted to get an order of 120 chickens. Two people started unboxing more chickens at the side of the shack, while a third was sliding chickens onto long metal spits, eight chickens on each six-foot pole.

"This is what everyone should be doing," said Alex, the deportation officer. "Anyone in a position of influence should just be doing things like this and making things happen."

I noticed a group of five women drinking beers at one of the tables and asked them where they were from. By chance, it turned out they were school cooks who had finished work already. They had cooked food not just for their students, but also seven hundred meals for their local community. It was yet another sign that the school feeding plan was working.

Along with the reopening of Maelo's, it looked like Puerto Rico's food recovery was developing quickly. This was no longer the crisis of a month ago, and we would have to adapt too. Maybe we could funnel FEMA money to local restaurants to help reach the areas of need. We could coordinate with local mayors and the local restaurants to feed the island and revive the economy at the same time. But first, we had to figure out what those areas of need really looked like.

We loaded up the back of the Jeep with tray after tray of delicious roast chicken, along with piles of rice and beans. The air inside was filled with the aroma of this most delicious chicken, and it was hard not to eat it as we drove back to the river. I don't think there has ever been a Homeland Security truck that smelled so good. The total cost for these five hundred chicken meals was

$1,130, which came out to just $2.26 per person. It was a great model for how we could move forward, if FEMA were open to thinking in a new way about expanding our reach across the island.

That was the good news. The bad news was that we couldn't reach the military Humvees. The satellite phone didn't work, or the team didn't pick up the call. We drove to the water treatment plant, but there wasn't any sign of them. We had no choice: we would take the chicken across the river ourselves. Only this time, I was determined to make it all the way across.

We started edging out into the middle of the river. To begin with, the footing was slippery but held, and the water was fast but low. As we got closer to the middle, the ground became rocky: the concrete ford had broken and it was hard to keep our balance. The water suddenly grew deep and turbulent, and we struggled not to get swept into the river. Some of the villagers met us halfway across and seemed to find the crossing much easier than we did. Perhaps it was all their practice over the last few weeks. They looked like they were walking on water to feed the hungry, like a modern-day Jesus.

On my third run, carrying a tray of hot rice, my foot hit a rock and I fell hard. My instincts kicked in and I held the rice above my head. Nothing would stop the food getting through, even if it meant that my knee took the full force of my stumble. I was soaked, and my knee felt like something had cracked. But the rice was still hot and good to eat. My two satellite phones did not survive the fall as well as the food. At least I had a third phone back in San Juan with Erin or Nate.

We carried on until all the trays of chicken and rice were on the other side of the river, where some locals met us in their pickup trucks to take us to the center of the village. "We're an island within the island," one of them told me. "We're surrounded by water."

There were few signs of cleanup in San Lorenzo. Houses that had collapsed in the hurricane simply lay where they fell. There was no sign of a working economy, and the villagers were happy to see us. One of the older men told me they had already distributed our earlier delivery of food, first to the elderly and then to the sick. Only this final round would go to younger, healthy people. People on the mainland liked to think of Puerto Ricans—especially those living in poor, rural parts of the island—as corrupt or criminal. But these people in San Lorenzo put their community first; they took care of the weakest before they looked after themselves.

"We have food already," said one of the men. "We have supplies. What we need are generators to keep ice cold. Can you get us one?"

At that point I knew we had reached something far more important than the other side of a treacherous river crossing. We had reached the point where the island was beginning to stand on its own, at least in terms of food and water. Life was far from normal here. But if San Lorenzo—supposedly the hardest-hit village in the area—was fine for food, then we could start planning to wind down our food relief. We could keep a few kitchens open to concentrate on places that needed the most help, but we didn't need to keep going at maximum volume for much longer. If we stayed too big for too long, we would crowd out all those small restaurants like Maelo's just at the point when they were trying to get back into business. And if we closed down too soon, we would be neglecting the people who still lived in terrible conditions, and still needed food relief in an economy that was struggling to recover without power and water.

On our way back to the river, we ran into the military Humvees again. They had arrived ahead of us, and moved on to another repair mission, before returning to see if they could track us down. We were only too happy to take their Humvees back across the river. They gunned over the potholes and we bounced across in less than a minute.

The next day we hit our all-time record for meals produced: 145,637 in a single day, from 16 kitchens, taking our total count to more than 1.5 million meals. Looking back at where we began, less than a month earlier, it was an amazing achievement. But it was now time to think about how we could shrink.

BEFORE MAKING ANY FINAL DECISIONS, I WANTED TO CHECK OUT THE most remote parts of our operation: the islands off the big island. I took another puddle-jumper from San Juan for the short flight to the small island of Culebra, which is actually closer to St. Thomas than to the main island of Puerto Rico. It looks like a Caribbean tourist paradise, with white beaches and palm trees, and it has a small airstrip that sits between steep hills. If it weren't for the fact that the U.S. Navy used it for bombing practice until 1975, Culebra might have been a household name. Instead, even in Puerto Rico, this seven-mile-wide island is easily overlooked.

A couple of staff were cleaning and running the car rental desk inside the tiny airport terminal. What they told me came as a surprise: the island had recovered more than the mainland, and *preferred* to be overlooked. They had power for the workday, from 6:00 a.m. to 6:00 p.m., although there were clearly no tourists to look after. There were more Jeeps for rent here than I had ever seen at San Juan's airport, so I paid for one and drove into town.

I found the government building easily: it had a dozen volunteers outside, handing out supplies to a handful of people who stopped for them. There were pallets of bottled water, and boxes upon boxes of MREs. Some of the military meals came from South Carolina; others from Ohio. There were boxes of crackers and potted meat from Georgia. A month after Maria, people were still handing out this plastic-wrapped version of manufactured food. No wonder there were few takers.

Our hot meals were coming by boat to the small jetty in front of the government center. As soon as the aluminum trays of rice and chicken started coming ashore, one of the volunteers shouted, "No more rations! Now we have real meals!" On board were my own volunteers, from Dame Un Bite, a Puerto Rican version of Seamless or GrubHub run by José Ortiz, which translates from Spanglish as Give Me a Bite. The red-shirted young volunteers were tireless in delivering up and down the stairs of apartment buildings with no power and no working elevators. They manned the front desk at El Choli, where the food orders came in, and here they were again, in this farthest corner of Puerto Rico. Just the sight of their shirts filled me with hope. They helped us serve sandwiches and plates of chicken and rice to the islanders, as someone nearby played "*Despacito*" on a portable speaker.

We returned to the airport and flew on to Vieques, where I wanted to see how life had moved on in the last two weeks. We drove down south to the Bili beach restaurant, where we cooked our food for the island and checked on their supplies. But what caught my eye was the restaurant next door, called Bananas, where the owner Kelly Soukup was reopening that day.

"We're losing money but we have to open for the staff and everybody else," she told me. "We're here for whoever is here. It is what it is. You can't just sit around at your house with no power, right? We're hoping people will come."

I loved her spirit and what it said about the recovery of restaurants and Puerto Rico. I gave her some cash to help her through, because we both knew there were no tourists to help her business. Not yet. "Buy whatever you need," I said.

"Are you sure?" she said. "Because I want to cry. No one has done that for me. We usually give, give, give."

In these moments I understood why Mercy Corps was doing the right thing. Every time I gave money, it seemed like the smart thing to do. Especially when you know the people and how they are going to use the money. It means a lot to a small business owner to get a small infusion of cash.

We drove back to Isabel Segunda to explore the food economy of the main town. There were more signs of life, in the middle of so many challenges. I noticed that a small pizzeria, called Mama Mia, was open for business so I went inside. The place was half full of customers and entirely full of conversation. There was food coming out of the kitchen even though they had no water. It wasn't exactly the healthiest conditions for cooking, but with pizza you can improvise more easily than with other types of food.

All these small details sent me one clear signal: it was time to wind down our operations. Our FEMA contract, signed just a few days ago, would only run for another few days. There was no sign of a third contract coming. Our last day at the arena was just a couple of days away, and the company managing the space was constantly asking for more money, with little notice, as it tried to make up for the collapse of its normal business. I didn't want to close down completely; Puerto Rico certainly was not ready for that. Without ongoing FEMA funding, we still needed to find where the communities of greatest need were.

As we prepared to leave El Choli, we notched up some record numbers that demonstrated what we were capable of, and pointed to what we could have done if there had been a real partnership with FEMA and the large nonprofits: we produced more than 146,000 meals a day for our last two days at the arena, crossing the total of 2 million meals on the day before we left. It was a heroic performance: in a single day we surpassed our entire cooking and sandwich preparation in our first ten days of food relief. The only disappointment came from FEMA itself. On the same day, they sent us an email to cease all production once the contract was fulfilled. "Our Mass Care unit has no further need and decided to cancel any further production/distribution of meal services," they wrote.

Either FEMA was still stunningly clueless about the needs of the island, or they had lined up another mass producer of meals and sandwiches. Perhaps the most likely explanation was a bit of both. In any case, they didn't want to work with us anymore.

Our 2 million meals compared pretty well to the grand total of inedible MREs distributed by the U.S. military: 9 million plastic packets of what was once food.[3] Taking together all the military supplies of MREs, including another 3 million sent to FEMA, that represented just two days of what FEMA thought the island needed to eat. If you're even more generous, and include the "shelf stable meals," otherwise known as cans, the U.S. military—with its vast budget and huge transport capabilities—supplied the island with around fourteen days of food. I suspected that most of those MREs were never distributed to people, but were actually getting stored in giant warehouses. Somewhere on the island is a massive stockpile, like the scene at the end of the movie *Raiders of the Lost Ark*.

On our last day at El Choli, I huddled with my food truck team and hugged them to say thank you. I gathered my cooks at the paella pans for one last song:

Voy subiendo, voy bajando: I go up, I go down.

Voy subiendo, voy bajando: I go up, I go down.

Tu vives como yo vivo, yo vivo cocinando: You live like me, I live by cooking.

Tu vives como yo vivo, yo vivo cocinando: You live like me, I live by cooking.

OUR NEW HEADQUARTERS WAS CLOSE TO THE SAN JUAN AIRPORT, AT THE Vivo Beach Club: a resort by the sea that was going to be closed for many months. But while its rooms and landscaping needed repairs, its huge kitchen was in great condition. We set up a new sandwich line in an empty conference room, taking care not to slop mayo on the carpeted floor. Outside my cooks could take a break gazing out to sea—or at the empty beach café. It didn't take long for us to get back to preparing thousands of meals a day.

We continued to get amazing public support. Lin-Manuel Miranda stopped by to see our kitchen and to thank everyone. He had already recorded a great song, "Almost Like Praying," to raise money for hurricane relief. The song featured the magical Taíno names of many of the island's municipalities, and was performed by Latino superstars including Jennifer Lopez, Gloria Estefan and Luis Fonsi. He called the song "a love letter to Puerto Rico" and warned against the compassion fatigue that made people ignore the suffering on the island.

In media interviews, Miranda acknowledged the historic parallel with his biggest hit: Hamilton himself appealed for outside help after a hurricane devastated the island of St. Croix in August 1772, not far from Puerto Rico.[4] Lin-Manuel's visit lifted the team's spirits, and he called our work "really inspiring"—which was really inspiring for all of us too. As he raised money for us with a Facebook campaign, he was happy to join with Erin, my chief of operations, as she rapped her way through the opening of his musical In The Heights. It was a dream come true for Erin and I was happy for her.

His visit was the kind of thing we held on to while others were less than supportive. Even though our formal relationship with them was over at the end of our contract, FEMA continued with its low-level sniping at me and our food relief operation. It was a petty dispute that spoke volumes about their attitudes and decision making in the middle of the biggest humanitarian crisis in the U.S. in living memory. A reporter at BuzzFeed asked FEMA about the end of our relationship, and Marty Bahamonde, who had been helpful to me on the phone, was dismissive of our work, portraying our food relief as some kind of marketing campaign for my restaurants.

He described me as a "colorful guy who gets a lot of exposure" and "a businessman looking for stuff to promote his business."

I won't lie. This kind of attack hurt me because it smeared the reputation of all the Chefs For Puerto Rico. The island was always prone to cynical rumors of corruption and FEMA was intentionally stirring them up. Why resort to such a deceitful caricature?

"He was very critical of us publicly and we were disappointed that he took that approach," Bahamonde told BuzzFeed. "We had a good working relationship, and we paid him a lot of money to do that work. It wasn't volunteer work—so we were disappointed in some of his public comments."[5]

But this explanation was also deceitful. FEMA didn't pay me a lot of money to work. They paid a lot of money to a nonprofit, World Central Kitchen, to reimburse us for the massive costs of producing millions of meals for Puerto Ricans. I wasn't paid a dime for my work, and my businesses were never compensated for my time or for the work of any of my chefs and executives who helped in Puerto Rico.

So what was riling them up, after they had split with us? And were those feelings the cause of the split in the first place?

The BuzzFeed comments were revenge for an article in *Time* magazine a month earlier, at the height of our dispute over the second contract, when FEMA had said there would be no further work and we would have to shut down early.

"People are hungry today. FEMA should be in the business of taking care of Americans in this minute," I told *Time*. "The American government has failed."[6]

Even with the distance of time, I stand by those comments today. There was no real sense of urgency at FEMA or inside the Trump administration. They failed to take adequate care of

American citizens struggling without food and water. I explained to *Time* that my first FEMA contract—for just twenty thousand meals a day for one week—was nowhere near enough for the island, or for us to cover our costs. "FEMA used me as a puppet to show that they were doing something," I told the magazine.

Perhaps I should have been more diplomatic in my public comments. But the situation was urgent and my public pressure seemed to be the only thing that was breaking through the government bureaucracy and the boredom of the media. Who else was going to speak up for the American citizens of Puerto Rico, and how else were they going to eat something other than plastic MREs for weeks on end?

FEMA admitted to BuzzFeed that bureaucratic red tape was at the heart of the problem. "The agency acknowledged that only his organization, World Central Kitchen, was able to offer hot meals on the island, but said FEMA could only offer Andrés short-term emergency contracts for two weeks at a time, not the longer-term contract he wanted because that would have to go through a competitive bid process due to federal procurement laws," BuzzFeed wrote, quoting an official saying there "was a frustration on his part in what he viewed as bureaucracy getting in the way."

That is something we can agree on. I was frustrated with the red tape that meant there could be no contract for several weeks while the bidding process took place. An emergency is an emergency, and a long contract negotiation does nothing for people without food and water. But I was happy with two-week contracts and I was even happier to compete with anyone on price. I just wanted to feed the people.

Besides, as we found out later, FEMA was writing huge contracts for people with no experience, no capacity to deliver meals and a track record of failure. Their arguments, like the comments to BuzzFeed, seemed more about personal pride and internal politics than anything else.

I texted Bahamonde and FEMA's tone changed. The next day they told the local newspapers that we had done a great job. I never knew if that bad story was the product of the press or the people of FEMA. Sometimes I wondered if we should stop talking to the media. But we would not have been taken so seriously without them.

At this point, FEMA itself had gone into the sandwich business, buying and distributing up to 100,000 sandwiches a day from the airline caterers. They told our friends at Homeland Security to stop supporting us, a demand our friends ignored, according to HSI agents who told me directly. FEMA had managed to ramp up just at the point where everyone needed to ramp down because the island was recovering enough on its own. By the time they managed to copy part of our food relief operation, it was too late to flood Puerto Rico with free food. Instead it was time to be smart, nimble and focused, qualities that FEMA had struggled to figure out.

A week later, in mid-November, the sandwich operation abruptly stopped. There was no transition and no announcement.

OUR SATELLITE KITCHENS AND FOOD TRUCKS WERE STILL SERVING RE-mote areas, as well as homes for the elderly, and finding they needed plenty of food. We were still making several thousand sandwiches each day, as we maintained a daily output of around thirty thousand meals.

I took supplies to the mountains in Utuado, where chef Jeremy Hansen from Spokane, Washington, had been cooking in a school kitchen for two weeks. He left his restaurants—Santé and Inland Pacific—because he was watching the crisis and wanted to help. "Being home, and seeing what was happening, I didn't feel there was a big enough response out here," he said. "I felt really compelled to figure it out and do whatever we can. These are things I've been wanting to do my whole life. It's a huge part of why I love to cook. It's life-changing in many ways. Look at the people here; they are amazing."

Hansen was producing a few hundred meals a day until he did a local radio interview about his community kitchen. The next day he served 4,480 meals as people lined up outside. He was also getting higher up into the mountains, where the people found it hard to get to a place like his kitchen, especially if they were poor and elderly.

But what about his restaurants back home? How were they doing? "They're going to be fine," he said. "I'm in contact with my friends there. It's routine. They're doing their thing. The really amazing thing is I almost don't care actually. I wish I could do stuff like this forever. It's better than anything."

Hansen was like my other chef partners: he was inspired to do this all over again in the next crisis. In that way, Puerto Rico was the start of a whole new movement that could change food relief for years to come. "I've been cooking for twenty-five years and trying to do whatever to get a Michelin star," he said. "But this is more meaningful than anything. I would love to keep doing things like this."

I drove on from Utuado to an even remoter town, where Erin heard there was someone in need of vital medical supplies. I found it hard to ignore these appeals, especially because I could also learn so much about the real situation in the interior. A forty-eight-year-old woman, Lilia Rivera, was seriously asthmatic and had run out of her basic inhalant, Advair. She had suffered a chemical spill at work and burned her lungs a few years earlier, and now she couldn't breathe comfortably without medical help. Lilia's town, Río Abajo, was entirely cut off from the rest of the mountainous area. A high concrete bridge spanning a ravine to the main road was washed away by the huge floods that swept down the mountain, three days after Maria made landfall.

On the mainland, you can get Advair at your local pharmacy

and pop it straight into a respirator. But the hurricane did not just cut Lilia off from her medicine. Her respirator needed electricity, and she was running it from her car, where the exhaust fumes were making her asthma worse. The coast guard had helped her earlier with new car batteries, but now she was out of Advair. To reach her, we had to park on the main road, climb down into the muddy riverbed, and walk across to the other side of the river where an improvised ladder had been tied to the broken concrete pillars supporting what was left of the forty-foot-high bridge. It looked like the remains of the bridge might collapse at any moment.

On the other side, we hitched a ride to Lilia's house farther up the hillside, where I gave her both the medicine she desperately needed and a few solar-powered lamps, including one that could also charge her phone. Her eyes popped out in amazement. She seemed frail: her breathing was heavy and she leaned on a walking stick. But the look of wonder and happiness on her face was unforgettable. "Thank you for all the joy you've brought and all the mouths you've filled," she said.

It started to rain hard, so we scrambled back down the hill to the bridge. We didn't know if a flash flood would come roaring down the ravine and wash away what was left of our route home. As the raindrops grew bigger and bigger, we ran to the other side and got in our car for the drive back to San Juan.

The night closed in, cloaking every home in darkness. Along the side of the ravine, I noticed a local bar open, in the dark, with a few people drinking cans of beer. We stopped to talk, and I gave them my remaining solar lamps to brighten up their evening. I also gave them some land crabs I bought earlier on the side of the road. Their faces lit up. I shared a beer with them and heard their stories about life in the dark, with no money, no clean water or good food. These moments were a special time to connect with the people of Puerto Rico, to understand their struggles, and what life was really like on the island. They hoped the power would return soon; they heard it was coming back, not too far away. But they were happy to survive: the river had washed away homes that were eight feet above the collapsed bridge, so they were drinking tonight as the lucky ones who made it.

That can of beer was like nothing else in the world. It tasted of something you couldn't find in San Juan or the mainland. It was the feeling of being alive.

WE WERE THERE TO HELP FEED THE PEOPLE. BUT AS THE RECOVERY dragged on slowly, we were increasingly seeing other urgent, medical needs, simply because we were traveling across the island and communicating so widely about our work. Around the same time we helped Lilia, I met Dr. Juan Del Río Martín, a Spanish transplant surgeon at the Auxilio Mutuo hospital in San Juan. The hospital was built and run by the Spanish more than a century ago, and it proudly displays its Spanish heritage with flags and emblems across its older buildings. One of these is home to the transplant center, where my friend is director, and he works with his team on saving the lives of patients with advanced cancer. Like other hospitals, the Auxilio Mutuo had been hit hard by Maria, with extensive flooding and loss of power. Now, a month after the hurricane, it was struggling with a second wave of emergencies: a severe shortage of anti-rejection drugs to prevent transplanted organs being rejected by patients' immune systems.

This was so far out of my expertise, I didn't know where to begin. How bad was the situation on the island if the doctors lacked such drugs? I could only think of emailing people with the very best connections, so I reached out to Dr. Jim Yong Kim, the president of the World Bank in Washington, and Dr. Paul Farmer at Harvard. Jim and Paul co-founded Partners in Health in Haiti, a pioneer in low-cost, community-based health care across the world. If anyone knew how to solve this problem, it was them.

They kindly jumped to it, putting me in touch with the senior staff at Boston Children's Hospital.

The children's hospital was committed to helping Puerto Rico, sending multiple shipments of medications to the island, in partnership with several other groups. For their first shipment, they assembled four thousand pounds of medicines and supplies. They told me they had thought they were well connected to the medical needs of the island, but had no idea about this critical shortage at the transplant center. Dr. Del Río and his team were down to their last units of epinephrine, or adrenaline, which was essential to their surgery. But the full list ran to ten medications they urgently required. The situation was far worse than normal, the surgeon told me, because the hurricane had led so many patients to delay treatment even longer than usual. Patients were coming in with enormous tumors because they were so busy struggling with all the other basics of survival, they delayed their own care.

The first shipment of essential transplant medicines arrived with the help of Direct Relief, an international medical nonprofit, just one week after I emailed Jim and Paul. Another, led by Jeffrey Akman, dean of medicine at George Washington University hospital in D.C., arrived a week later. I am an adjunct professor at the university and know the president and faculty well.

When you see a humanitarian disaster, you can always help, even if this isn't something you know anything about. The fierce urgency of now means you act today, not tomorrow.

ALMOST TWO MONTHS AFTER MARIA, THERE WERE STILL, UNBELIEVABLY, parts of the island that looked exactly like the hurricane had ripped through them yesterday. At Humacao's small airport, on the island's east coast, it was impossible to drive down the access road. It was still blocked by power and telephone poles and cables. For that matter, it was hard to *walk* down the road. You could only hope the electricity wasn't restored in some places. A small hangar at the side of the airstrip bore witness to the force of nature: planes were casually hurled into the building and a helicopter was destroyed nearly beyond recognition: you could barely make out its shell. A few minutes away, cars were lined up for several blocks outside a local supermarket. They were expecting a delivery of water and food, so all normal life ceased while people waited for their chance to buy the basics of life.

In Punta Santiago, where Maria first landed, the area was like a demolition site. Gas stations were flattened and homes abandoned. One elderly resident, Don Alfonso, lived in something that was both flattened and standing. You could see into his kitchen from the street because the entire front of his home had disappeared, along with most of his roof. All that was left was his bedroom, and the plastic over his head was leaking. After so many weeks essentially living outdoors, in what remained of his home, there was still no sign of a FEMA tarp. We promised to buy one ourselves to help him stay dry.

We started delivering in this neighborhood as soon as we saw photos on social media of the sign the townspeople painted on the road at the end of Don Alfonso's street: SOS Necesitamos Agua/Comida!! We Need Water/Food!! Imagine how desperate you need to feel to paint that message in the road, in the hope that a helicopter or plane sees it and comes to your rescue. We had a food truck here three hours after we first saw that plea for help. Soon we started deliveries of hot food to a nearby community center, but coming here always made me feel stupid. How could we all complain about our daily troubles when these people suffered so much, yet so quietly?

One block away was a small nursing home that saw the worst of the storm surge. Within minutes the water quickly rose to four feet, threatening the lives of the five elderly patients inside. A neighbor broke open a window, tied one end of a rope around a concrete column at his house and attached the other end to each patient, to stop them from being swept away. He pulled the patients out to a ladder where they could climb to his second-floor balcony.

Several weeks after their escape, these frail Americans were still sleeping on mattresses that had been soaked in seawater. Their beds were broken and black mold was growing on them, as it was inside many of the homes here. The nursing home manager Violeta Guerrera had no money to replace the beds, so I promised to buy them myself with money donated by my friends, the Milstein family, who have been coming to Puerto Rico for many years. I asked them if it was OK to use the money for something other than food and water, and they said yes: I could use it in any way that was necessary.

This was America and people didn't want our pity; they wanted our respect. And the way you show respect is to provide the things they need, when they need them.

WE BEGAN TO FIGURE OUT A WAY TO RAMP DOWN FURTHER. THANKSGIV-ing was only a couple of weeks away, and we wanted to make one more big push to feed our fellow Americans. What better way to tell them America cared than with a plate of turkey? We also wanted to thank all our volunteers and partners with a big Thanksgiving dinner. Soon after, we would wind down our numbers to a low rate in just a handful of places.

I visited Eva Bolivar in Vieques, the owner of Bili restaurant, and we agreed to reduce our daily meals to five hundred through the end of the month. She still needed help making her $6,000 mortgage each month, through the end of the year, and we promised to help get her restaurant back on its feet. Vieques would continue to struggle with its recovery, as gas supplies would stop abruptly and the islanders were forced to line up for hours waiting for the next shipment.

Other kitchens would stay open longer. At the Hilton resort

in Ponce, in the south, chef Ventura Vivoni was producing thousands of meals in a corner of the hotel's vast kitchen to deliver in the mountains north of the city. People didn't want to waste their precious gas so the food needed to travel to them. I promised Vivoni we would support him, even as we scaled back our production.

As Thanksgiving approached, we neared an epic milestone: 3 million meals in the two months since the hurricane. Before preparing a turkey dinner for one thousand volunteers and their families, I wanted to see again some of the hardest-hit corners of the island, so I returned to Punta Santiago. The neighborhood had barely changed. The gas station was still destroyed and the electric poles, made of reinforced concrete, were still lying on the sidewalk, snapped like so many pencils. Outside the nursing home there was a pile of moldy mattresses, smelling like a blocked toilet. But inside, the patients were finally sleeping on something we all could recognize as a bed.

On our drive back to San Juan, the road was blocked by a party outside a small Catholic church I had never noticed before. There was an inflatable Nativity scene on the roof, close to the sign above the door saying *Parroquia Nuestra Senora del Carmen*. Outside there was a band getting ready to play on a stage, and a giant barbecue of delicious chicken skewers and *bacalaito* fritters with salt cod. People were dancing and young families were eating together. Inside, a Puerto Rican flag hung behind the giant crucifix. Below, stuck to the wall, were handwritten signs that told the story of how this community had survived. *Paz*, peace. *Perserverancia*, perseverance. *Humanidad*, humanity. *Vida*, life.

Pastor José Colón told me how the floodwater had risen above the pews, but you wouldn't have known that from the freshly painted white walls. Behind the church, under blue tarps, he stored pallets and boxes of food to distribute to his community.

The food came from private donations; he had asked for government help, but nothing came.

I told him what we had done to feed the island, and he asked me a simple question.

"Why did you come to Puerto Rico? Where did the idea come from?"

I was, for once, at a loss for words.

"I don't know. I didn't have a plan. My idea was just to feed the people."

EPILOGUE

WE HAD ALMOST 3 MILLION REASONS TO GIVE THANKS. THREE MILLION meals prepared for hungry Puerto Ricans by so many chefs—from the island and the mainland—and no less than 20,000 volunteers working across twenty-four kitchens and seven food trucks. We wanted to say thank-you with a special Thanksgiving dinner—that uniquely American meal—for our cooks, our partners and as many volunteers as we could fit around several long tables inside a conference room at the Vivo Beach Club.

I introduced Eliomar Santana, from the church high up in the mountains of Naguabo, explaining how I heard about a pastor who wanted to cook and of course I would have said no, if anyone had asked me first. Now Eliomar had cooked tens of thousands of meals, and inspired us all. He said grace, giving thanks for us all, and we hugged.

I warned everyone to be careful. "Okay, I cut myself already," I said, "so please only let someone cut the turkey if they know how to use a knife."

I thanked as many chefs as I could name and all of those I couldn't. I thanked our volunteers and our food suppliers, especially

José Santiago. And I gave a huge thanks to José Enrique and his sister Karla for starting the whole story.

"You know what happened? We needed a restaurant and they gave us their restaurant," I said. "We needed a car park, and he got us a car park. And we began cooking *sancocho*, the best *sancocho* in the history of mankind. And we began making sandwiches, the best sandwiches with mayo in the history of humankind. And then in the parking garage we began getting food trucks. And we began getting paella pans. Paella pans! A crazy guy called Manolo came from Miami wanting to cook rice. He and his team have done hundreds of thousands of *arroz con pollo*, day in and day out. Chefs For Puerto Rico were so many people, but we needed angels, and the angels we had were food trucks. They came to us and we didn't have gasoline, but we traded gasoline for food. And that's good because we were feeding people."

Above all, I wanted to thank the people of Puerto Rico: all those struggling but selfless communities who told us there were others who needed our help more than they did.

"When you find people that generous, that's when you really see the real beauty and meaning of the American phrase *We the People*. It's not *I the Person*. That is the island of Puerto Rico. Thank you to Puerto Rico, thank you to the chefs for being part of this and for feeding so many people. *Viva Puerto Rico. Puerto Rico se levanta!*"

We finished the only way I knew how: with the loudest singing of our anthem:

Voy subiendo, voy bajando
Voy subiendo, voy bajando
Tu vives como yo vivo, yo vivo cocinando
Tu vives como yo vivo, yo vivo cocinando
You live like me, I live by cooking.

HOW DID A BUNCH OF CHEFS AND VOLUNTEERS ACHIEVE SO MUCH IN such a short time? Because we weren't just a bunch of cooks with great knife skills and the ability to conjure up great flavors. Restaurants are complicated businesses, and a great chef needs to be a great manager—not just of people, but of orders, supplies and inventory. If you can't get the management right, it doesn't matter how good a chef you are: your restaurant will fail. Those skills, it turns out, are incredibly useful in a disaster zone. David Thomas, executive chef at my Bazaar restaurants, was a perfect example of the kind of logistical genius you need to run these relief operations. In his day job, he oversees four $10 million restaurants, managing daily orders of 10,000 or even 20,000 items. Still, Puerto Rico was something else. "The sheer volume of things coming in the door was crazy," he said. "And Puerto Rico doesn't have the most reliable food distribution services. Unreliability was a big challenge. Everybody wants to say, 'Yes, no problem, we will get the order.' But then it doesn't show up, which means at some point there's a complete hole in production, which can't happen."

Chefs understand how to create order out of chaos, just as they know how to control the fire to cook great meals. There were lots of moments when we didn't know what to do in the early days. The conversation would go like this: *What the fuck do we do next? Okay, let's keep cooking. That's a good plan!*

Harvard Business Review ran an article recently about embracing complexity, citing the great example of an ant colony.[1] Each ant works with local information, and has no big picture of what's going on. It has no plan, and no obvious leadership, yet together the colony achieves incredible feats of organization and engineering. What we did was embrace complexity every single second. Not planning, not meeting, just improvising. The old school wants you to plan, but we needed to feed the people. We were sending food trucks to those who were fainting in line for food because it was two weeks after the hurricane and not even MREs had reached them. I didn't call an expert in painting or the history of the ninth century. I called the experts on how you get food to the people in very little time and on very little budget: cooks and kitchens and suppliers.

If we had a plan, it was to be united to achieve as much as possible. With El Choli, we were the biggest restaurant in the world. Period. And if we put in all the kitchens and the food trucks, we were the biggest restaurant company built in the shortest amount of time. How many restaurant companies go from one restaurant to sixteen in less than two weeks, unfunded? Everyone kept saying we needed to have a plan, but we never organized. How many days are you going to organize when people are going hungry? People were eating roots. American citizens eating inedible roots. This was not a far-away country on another continent. This was American soil. That passion to help our fellow Americans was a big reason why we stayed united against the odds, and why we cooked for so many people.

HUNGER AND THIRST ARE HARD TO SEE, ESPECIALLY IN A DISASTER RE-covery situation like Puerto Rico. You don't even realize what the problems are until you go into the community and talk to people.

We gave them food that was prepared and cooked earlier the same day. We made it locally to bring economic activity back, paying chefs and hiring food trucks, to create the conditions so people could feed themselves. Along the way, we were restoring pride and creating the conditions so that law and order could return. People behave strangely when they are hungry: they will break into stores to steal food, and risk getting arrested or shot, if there is no other way to feed their families. That's why my friends in Homeland Security Investigations told me that delivering our food made their job so much easier. It's far better to approach someone with a sandwich than a gun, no matter if you're carrying a gun anyway.

Governments and nonprofits exist to serve the people: What point is disaster relief if you don't care for the people who need that relief? There can be no greater priority than food and water, even though FEMA officials told me otherwise. That's why someone needs to be held to account for the delivery of food and water in a crisis: a food tsar who can cut through the bureaucracy to save and rebuild lives. We cannot leave this vital task to the elderly volunteers of the Southern Baptist Convention alone.

With food as a priority, we should go into any disaster area with a pre-prepared plan of action. We know where the hurricanes, tornadoes and earthquakes strike. We know there will be more of them as the climate continues to change before our eyes. And we know that the first few days after a catastrophe are critical in terms of food and water. We can stockpile supplies, identify kitchens, and alert relief workers to be on standby ahead of a predictable disaster like a hurricane. We can distribute emergency communications systems like satellite phones, in case the cell phone network goes down. We need emergency feeding teams ready to enter these disaster zones within twenty-four hours, just as we have search-and-rescue teams ready to pull people out of earthquake rubble. You don't start work after the emergency happens. We need a network of Food First Responders, or FFRs.

No two areas are alike, and you can never plan for everything. But you need to prepare for some things when food is a priority. You need filters to purify whatever supplies of water you can find, instead of trying to ship giant numbers of plastic bottles. You need generators, especially for the big arenas. In the San Juan convention center, half of the kitchen didn't work. People said it was a private business, but in an emergency it belongs to the people of Puerto Rico. In emergencies, cities should have the power to take back these giant kitchens temporarily, to serve the people. They should be feeding lots of people because the kitchens are big enough to do so. Education departments should have the power and budget to expand the cooking in their school kitchens, so they can serve their communities. Frankly, I'm amazed these plans are not in place already. With that kind of thinking, we could have taken care of the island with half of the people at FEMA.

We could also innovate around the delivery of food. Instead of buying and trashing millions of throwaway plates and cutlery, we could hand out reusable plates and cutlery at the start of a disaster. Believe me, people will take care of them if they know the emergency plate and cutlery guarantees them a hot meal the following day.

WHAT WENT WRONG WITH THE DISASTER RELIEF IN PUERTO RICO?

The simple answer is: most of it. When you see such epic failures, you realize these are systemic problems. The system failed from the top to the bottom, at every level of government, from federal agencies to nonprofit charities. That's not because there were bad people with malicious intent. I said from the start: there were many good people trying to do good things. But their thinking and their organization was all wrong, and the results were not just inefficient, but deadly.

This isn't just my view: the thoughts of a chef who didn't complete high school. Refugees International made its own assessment of the disaster relief in Puerto Rico two months after the hurricane, in its first mission inside the United States in its thirty-eight-year history. Based on its huge international experience, it wanted to compare best practices overseas with what was going on in America. The conclusions, published at the time we were winding down our cooking, were shockingly bad.

"Our team encountered a response by federal and Puerto Rican authorities that was still largely uncoordinated and poorly implemented and that was prolonging the humanitarian emergency on the ground," they said. "While food and bottled water are now widely available and hospitals and clinics back up and running, thousands of people still lack sustainable access to potable water and electricity and dry, safe places to sleep."

The group recommended stronger leadership and coordination with local officials and community groups, much better information and communication, targeting the most vulnerable for urgent help, and applying international best practices for future disasters.

"Unfortunately, the response to the catastrophic disaster in Puerto Rico lacked the requisite leadership from the highest levels of the U.S. government necessary to support a more effective, timely response by FEMA," they concluded. "The response remains poorly coordinated and lacks transparency. Disaster survivors continue to face horrendous living conditions and lack information on whether or when they will be assisted."[2]

There is no substitute for leadership at the top, starting inside the White House, with the president of the United States.

But Trump's obvious failures do not tell the full story. There is a kind of group thinking that leads to these systemic failures. After all, it's not as if the government and NGO officials didn't plan for a disaster, didn't hold endless meetings and didn't send long emails to giant contact lists. They tried hard, and they failed hard.

However, their attitudes were modeled on a top-down approach that was divorced from reality, and they were not alone in thinking that was the best way to deal with a disaster. Top-down approaches are favored in disaster zones but they are based on fear, and don't succeed as a result. As Erik Auf der Heide, the disaster management expert, writes, "The unfounded belief that people in disasters will panic or become unusually dependent on authorities for help may be one reason why disaster planners and emergency authorities often rely on a 'command-and-control' model as the basis of their response . . . This model presumes that strong, central, paramilitary-like leadership can overcome the problems posed by a dysfunctional public suffering from the effects of a disaster . . . Authorities may develop elaborate plans outlining how they will direct disaster response, only to find that members of the public, unaware of these plans, have taken actions on their own."[3]

As for the fear and anarchy, Puerto Rico demonstrated that those widespread expectations were entirely misplaced. Jorge Uribarri, the Assistant Special Agent in Charge of the first HSI team we partnered with, told us that his entire team of agents saw only one potentially dangerous situation during thousands of missions over many months. That was a single supermarket looting on the poor eastern tip of the island on the day the hurricane landed. Even then—in the most impoverished, worst-hit corners of the island—his agents saw no violence.

What works in a disaster is localized decision-making. That was clear from the contrast between the private and public sector responses to Hurricane Katrina in New Orleans, as Atul Gawande points out. "I talk about Katrina because you saw two kinds of checklists in action," he told a Harvard audience. "One is the kind of set of protocols that FEMA had in place, which was all about centralizing control. And in that instance where the protocols really dictated what people out at the periphery had to do, right down to their most nitty-gritty decisions, the thinking could not keep up with the scale of this disaster and its complexity. It just wasn't something FEMA had dealt with before. And the result was total failure. People doing the equivalent of standing at the bridges saying, this water delivery isn't on my list.

"By contrast, Walmart handled the situation by first telling their front-line managers in the stores, 'This is a situation that's beyond anything we've dealt with. You are all going to be working above your pay grade. We have a few key things that we need you to do, though. Number one, do everything you can to save people using your judgment about what is best. But second, communicate whatever you're doing on a daily basis to us, at the center, in their command center, and also, to one another.' Because if chaos wasn't to develop in the ways that different stores were handling things, they needed communication. So they focused on communication. And as a result, really great ideas spread immensely quickly. The stores became the place that opened up three pharmacies to make sure that local residents had access to medications that were critical. They were the quickest in moving emergency equipment to fire and police, whatever they had available in the stores. And they got water into the city two days before FEMA did.

"It had to do with the idea that under situations of complexity, you want to distribute power out to the periphery as much as possible. But then, encourage those teams at the edges to understand the individuals are fallible. But teams of people are more likely to get the better results."[4]

Even by the standards of top-down management, the government failed. The Pentagon—the epitome of a top-down organization—failed to do its normal job of pre-positioning assets ahead of Maria. After Hurricane Irma passed through, the defense chiefs sent their resources home rather than responding to Maria's explosive threat. They didn't even appoint an on-scene commander, Brigadier General Richard Kim, until ten days after the hurricane left Puerto Rico.[5] When the military showed up in Puerto Rico, they were simply too little, too late. "We're replaying a scene from Katrina," said Army Lieutenant General Russel Honoré, the man who saved New Orleans in 2005. "We started moving about four days too late."

It was no coincidence that Walmart was forced to throw out tons of food in Puerto Rico because the federal government failed to respond to its pleas for emergency fuel supplies to keep its food fresh. Pressure from Puerto Rican officials and members of Congress made no difference. FEMA simply chose to say and do nothing, even when Walmart executives said it would take weeks to replenish their supplies.[6]

The island's government was not much more impactful. Leaders like Julia Keleher struggled to communicate with her own schools. The mayor of San Juan was more present on television than government headquarters. The governor seemed paralyzed by internal politics when it came to solving the water crisis for his own desperate people. They were so overwhelmed, by the disaster and by long-running challenges, that they could not lead effectively.

As for FEMA itself, the agency remained stuck in a state of denial, unable to see its own disastrous failures. Six months after Maria, Brock Long, the FEMA administrator, testified to members of Congress that the massive $156 million food contract with Tribute that failed was in fact no big deal. "Out of 2,000 contracts, only three were canceled," he explained. "Tribute being one of them." Long did not say how many large contracts his agency negotiated for feeding Puerto Rico, but it's unlikely there were any bigger ones.

"The bottom line is my agency made a herculean effort to put food and water in every area," he continued later in the same hearing. "It's more complex, and it's not going to move as fast, when you are talking about an island jurisdiction, and the airports are completely blown out and the ports are blown out."[7] FEMA's "herculean effort" was news to any of us who were in fact putting food and water in all 78 municipalities. It was also plainly wrong to say the airports and ports were destroyed. Either Long was trying to deceive members of Congress or he was incredibly ignorant of the facts on the ground. Six months after the disaster, FEMA owed itself and the American people a better accounting of what went so catastrophically wrong in Puerto Rico.

The giant nonprofit sector suffered from similar failures. At their worst, they are as bureaucratic and political as the government, with an overriding mission of raising money to support their giant staff. They suffer from a savior complex, believing they know best how to govern the locals, rather than building up the local leaders and the local economy. Oxfam's sex abuse in Haiti was the very worst example of that mind-set of cultural and personal superiority, combined with an immoral lack of human respect and decency.[8]

This is the moment to demand answers from the big NGOs. If we keep giving money to them without results, we are doomed to fail again. We have to challenge their thinking to have a return on our investment. As an American and as a citizen of the world, I felt let down by the lack of readiness and preparedness of the Red Cross and the Salvation Army in Puerto Rico. They could have done better, but instead it was just business as usual. It was disaster boredom. It was just another crisis.

After Hurricane Sandy, the failures of the top-down approach of the Red Cross were obvious to its own leaders.[9] Richard Rieckenberg, one of the Red Cross heads of mass care, explained in a letter to the Red Cross vice president soon after the disaster, "As a matter of political expediency, we became committed to creating the illusion of providing mass care rather than the reality. At the level I was dealing with, this was done in a very deliberate and cynical manner. We became focused on making 'the numbers look good' and in 'showing a presence.' I was in an interesting position as the Mass Care Planner. I was not asked to plan. Rather, I had plans given to me which I was expected to endorse. Some were absurd."[10]

Some of the smaller nonprofits are leading the way, accepting the need to change. Mercy Corps realized that cash, food and water filters were key in Puerto Rico. They coupled those insights with great intelligence about the facts on the ground of who needed help the most, and in what areas. "We can't keep doing humanitarian aid the way we did in the 1990s in the Balkans with all the convoys of food," said Javier Alvarez. "We have to tap into new resources, private business, universities and social media. It's got to be much, much faster. We can't be sitting on the problems."

Even at the Red Cross, there are signs that change is coming. Brad Kieserman says it is his personal view that the organization needs to plan for disasters in a very different world. "If you are exchanging business cards at a time of a disaster, you're too late," he says. "You've got to be engaged beforehand. Do we as an emergency management community need to open the aperture further? We do. We are seeing a world in which the people are aging, the infrastructure is aging, the threats are becoming more aggressive and relentless. And I don't think anybody can deny the number of significant weather events has increased in size and scope, but also frequency." After all the floods, wildfires and hurricanes of the past year, Kieserman says these weather events are no longer outliers: "I don't think those things are an anomaly. I think they are a bellwether. I also don't think they are like what we've seen before on a regular basis. That's an important distinction. So we've already begun to plan differently. Because the ways that we planned, and the frequency and intensity for which we planned, are no longer the reality we face."

YES, WE NEED TO FOCUS ON THE LIBERATION OF THE RECEIVER. YES, WE need to meet the needs where they are. But I saw something profound change among the volunteers who worked so hard to feed an island. It was difficult work, for sure. Yet it was also one of the most fulfilling things they had ever done. Cooking changed them as much it changed the community. Making these meals renewed their values and identity: it reminded them who we are as Americans and what we stand for. In that sense, our country and our leaders would do well to reform disaster relief at home and internationally. We can find ourselves in these crises, as individuals and as a country.

This was a tough assignment for my team, separated from their families for a long time, sweating through the chaos of a tropical disaster zone. Our food relief operation was a huge challenge for each and every one of them. "It was emotionally and physically draining," said David Thomas. Still, in the middle of this intense recovery effort, they experienced some of the biggest achievements of their lives in the relationships they built with one another. Thomas will never forget the first time he visited the church in Naguabo. "They are all doing this because they honestly, truly want to help," he said. "It's like SEAL Team 6 going into a wartime situation," he explained. "Only these people can understand what we're going through. We're now friends for life."

For our Puerto Rican chefs, the extreme challenges of the recovery helped restore their sense of hope in their communities. "I have new faith in people," said Ricardo Rivera Badía of El Churry. "You see people here, and they've been here for weeks not getting paid. But they are as inspired and happy as the first day. They got engaged with the situation and empowered through it. They took it seriously and that definitely has created in me a new faith in people, to be candid. There are lots of stories of people we have offered money to and they have refused it because they are doing things from the bottom of their heart."

There are also lots of people who were ready to quit the island but instead found their commitment reborn. Ginny Piñero, who managed our order-taking and organized us from our earliest days, said she found purpose in our food relief efforts. She had been planning to leave Puerto Rico before the hurricane hit, as her son was going to college on the mainland, and her career needed a new chapter. Then her life changed. "I feel tired, but I feel satisfied," she said. "I feel complete." She had spent much of the last year working on a doomed campaign for governor by David Bernier, helping to shape the candidate's policies in the hope of serving as his representative on the island's fiscal board. When she talked to him in the middle of our food relief operation, she told him how her life was going. "If you were the governor, I was going to be on the fiscal board," she said. "I don't have to deal with the fiscal board now. I have to deal with the whole island." After her experience, she decided she needed to go into public service of some kind, like her great grandfather, who was governor of the island in the 1940s. "That is me," she says. "That is who I am. This is my legacy to look after the people. It's confirmed I need to make a whole turn in my career. I'm going to keep on working with non-profits."

As for Puerto Rico, my friends don't expect anything to come back until 2019. Even then, Puerto Rico needs much longer term help to restructure its debts and rebuild its infrastructure. "I just don't see people coming and wanting to spend their hard-earned money immediately," said Wilo Benet, one of our original Chefs For Puerto Rico, whose great Pikayo restaurant was closed with the rest of the Condado Plaza Hilton in San Juan. His staff scattered across the United States, in Texas, Florida and Illinois, and he was hoping to bring them back when the time was right. Like many others, he continued to pay some employees who needed help for as long as he could, even without a functioning business. "I'm looking forward to when we can get together again," he said.

After the wars in Iraq and Afghanistan, politicians liked to say they wanted to see nation-building at home. Well, Puerto Rico is home and it needs to be rebuilt. If we can find billions to do that in the Middle East and central Asia, we can surely do that on American soil in the Caribbean. Nobody should be talking about FEMA pulling out or the disaster being over within a few months of the hurricane. It wasn't, and we were still cooking hot meals in Puerto Rico many months after we wound down our huge kitchens, when the recovery was supposed to be well under way. The line between disaster relief and poverty relief is blurred: the hunger of the elderly and the sick has been worsened by the economic collapse after Maria and the continued lack of clean water and reliable power. The latest estimates put the death toll in Puerto Rico at 4,645.[11] That is more than the number of people who died in the terrorist attacks on 9/11, and more than double the death toll after Hurricane Katrina flooded New Orleans. The best way to honor those American lives is to rebuild the place they left behind.

THERE SADLY IS ALWAYS ANOTHER DISASTER AROUND THE CORNER. While most of my team was still in Puerto Rico, we watched from a distance as wildfires destroyed the beautiful wine country of Napa Valley. My friend Guy Fieri started cooking thousands of meals for evacuees and first responders, saying we were his inspiration. "If this guy is able to go to a city that doesn't have power, doesn't have running water, and he's able to start feeding thousands of people, we gotta figure this out," the chef told a local radio station.[12]

In turn, when we saw the wildfires in Southern California in early December, we knew we had to get back in the kitchen. I called Nate and he said he was on the way before I even asked him to go. Working with my mentor Robert Egger at his new community operation, L.A. Kitchen, we began to build a new team of ten chefs under Jason Collis of Plated Events, and Tim Kilcoyne of Scratch in Ventura. Partnered with the Red Cross, we cooked fifty thousand meals for first responders in December and January, first for the wildfires and then the mudslides. Chef David Chang of Momofuku fame dropped by to help cook as the word spread about what we were doing. Our California operations demonstrated how this new model of food relief can work across different places and crises, as long as you are ready to adapt and work with local experts and partners.

A lot of people praise what I've done, but that isn't what should be praised. What I'm amazed about is that—without any infrastructure and without any readiness—we were able to be more ready and adaptable than organizations that specialize in emergencies. Imagine what we could have done with the right partners who wanted us to succeed. Imagine what we could do with the right technology: with digital maps showing in real time where food is needed and getting delivered. Imagine if we could do the same with medicines, seeing where drugs are stored and needed, like a digital pharmacy.

Our greatest achievement was not that we did all this. It's that we were able to grow from an organization that wasn't supposed to be there, to an organization that had better intelligence, supplies and purpose than the people who claimed they were experts. Now is the time for those disaster experts to work with the food experts to build a new model of disaster relief that is effective and efficient, driven by the right priorities.

That wasn't what we set out to do. We didn't want to fight with the bureaucracy or restructure government. All we wanted to do was to feed the people. But when you start with a simple goal, you learn you can achieve the impossible.

You discover, before long, that you can actually feed an island.

ACKNOWLEDGMENTS

IT WASN'T BY CHANCE THAT WE CALLED THIS BOOK WE FED AN ISLAND. I am humbled by the amount of personal recognition for my work in Puerto Rico. Unfortunately it seems that every movement needs a face. But I know, and you all should know, that it took many people, without expecting anything in return, who gave their best to fulfill a dream at a time when people were in need of a basic meal. Americans came together to form an army that grew by the hundreds every day. People recognized a moment to serve—an important moment—and that's what they did.

Here are my acknowledgments of the people who fed the island. I'm extremely sorry if you're not on this list. If we forgot you, please know that we are grateful to you forever.

To Nate Mook, who in just a few hours was sitting next to me on the plane and became the best friend anybody can have in a moment like this: unselfish, giving, calm in the chaos. Nate helped build the operations from the beginning and did everything from social media to being my trusted adviser.

To José Enrique Montes, his sister Karla, his father José Antonio Montes and the team from his restaurant José Enrique in Santurce. His restaurant became our fort from our very first days on the island after Maria. And to the most loyal, hard-working team from his restaurant: Ivan Lugo, Janiliss Hernández, Miguel Díaz, Erick Rondon, Daniel Sánchez, Denise Ortiz, and Andres Rossy, who are from now on my brothers and sisters forever.

To my chefs for Puerto Rico: Wilo Benet from Pikayo, who gave his best from the beginning, as well as amazing *pastelón de carne*; José Santaella from Santaella, who gave us his entire walk-in full of food; Enrique Piñeiro from Mesa 364, who became our second kitchen and the inspiration that we could open many more kitchens across the island; Victor Rosado, who has traveled with my ThinkFoodGroup from Mi Casa to Bazaar Mar; Manolo and Juan Martinez from Paella y Algo Mas, who worked harder and smiled more than anyone else, delivering thousands of extra meals wherever and whenever we needed them; Ventura Vivoni of Vida Ventura, a talented chef in the heart of Puerto Rico who worked to feed people when there were no cameras or reporters, going up hills and down valleys to deliver meals every day; Eva Bolivar of Bili Restaurant, who rose to the occasion even after her restaurant was badly damaged, partnering with us to cook thousands of meals on the beautiful island of Vieques; Carlos Perez of El Block, a rock on Vieques next to Eva, who made sure the elderly were never forgotten; Pastor Eliomar Santana of Iglesia de Jesus-cristo Monte Moriah in Naguabo, who is my favorite pastor-chef in the world, and who gave us hope that anything is possible with faith and love for each other.

To my friends and partners at the Instituto de Banca y Comercio, especially Gonçal Bonmati and Michael Bannett: from the earliest days, we began dreaming that we could activate your schools and feed everyone, and that dream became a reality thanks to your perseverance and willingness to help the people of Puerto Rico. I know the IBC kitchens will be there to support another tragic event, God forbid.

To my team: Erin Schrode, my American sister, who came to

support Nate and me without a clear idea of what we were talking about, but very quickly became the driving force of many of the things we did, and who stayed on the island for many more months, committed to feeding the people in need; David Thomas, one of the best chefs I know, a friend and a brother, who showed more passion than anyone for what was happening in Puerto Rico; David Strong, who I'm honored to say has helped me open many ventures, and who only had to look into my eyes to understand what I needed and what was needed to get this done; Jesus Antonio Serrano Pabon, my pastry chef from Minibar and his wife Alexsandra Ortiz; Michael Rolon; Ricardo Heredia; Tito Vargas; and to Ruben Garcia and Aitor Zabala, who are my left and right in creativity.

To Kimberly Grant, the best CEO on the planet, who could see from a distance that I was getting under water as our operations were growing bigger. In the shadows, she was able to make sure we got finance, accounting, lawyers and all the many things that you need to have in place as the operation exploded.

To Michael Doneff, Margaret Chaffee, Daniel Serrano, Stephanie Salvador and Satchel Kaplan-Allen, who kept everyone in the loop, working behind the scenes to make us successful and make me look good every day.

To Scott Sinder and his wife Jodie Kelley for always being there for me, every moment I need friends and lawyers. Much of my success in life I owe to them.

So many people from my company donated huge amounts of personal time and effort making sure that World Central Kitchen and I would not fail. We always said that TFG would change the world through the power of food, but I never imagined that through this company we would have the opportunity to serve the world as one. Once a TFGer, always a TFGer.

My special thanks to the chefs who came with me to Houston: Charisse Dickens, Victor Albisu and Faiz Ally.

To the core team in Puerto Rico: Ginny Piñero for becoming not only our order-taker but our connection to the island, and for often being the first to arrive in the morning and the last to leave at night. Also to Jennifer Herrera, Alejandro Perez, Alejandro Torres, Alejandro Umpierre, Andres Acosta.

To our chefs from Bon Appétit: Blas Baldepina, David Apthorpe, Khori and Juana Thomas, Ty Paup, and Karla Hoyos.

To all our chefs, cooks and volunteers: Carlos Carillo, Christian Carbonell, Christopher Knapp, Fernando Concepción, Eric Luis Lopez, Israel Rodriguez, Ivonne Rios Mejia, Javier Mercado, Camile Mercado, José Rios, José Ruben Martinez, Kalych Padro, Lymari Figueroa, Mariana Carbonell, Giselle "Ñaña" Villa, Dilka Benitez, Rosela Angela and Yolaida in Loiza, and Chef Jeremy Hansen.

To our *arroz* operation: Robert Espina, Stephanie Ortiz, Wandy Ortiz, Javier Liriano Feliciano, Oscar Maldonado, Yamil Lopez, Rawi Leafar Yuri Disla, Javier Liriano Feliciano, Juan Torres.

And to the more than 20,000 volunteers without whom we could never have fed an island.

To Rafa Herrera: As I was traveling the island, with multiple phones, I needed somebody to take me around but I never could have expected to make a friend like Rafa. He did far more than drive my car: he was the ultimate insider who gave me daily situation reports on the island.

To our angels, the food trucks of Chefs For Puerto Rico: Xoimar Manning, Michael Sauri and Alondra Sauri (Yummy Dumplings); Yareli Manning and José Gonzalez (Meatball Kitchen); Ricardo and Luz Rivera Badía (El Churry); Mariana Lima Limoso (Acai On The Go); José Ortiz and Team Develop (Dame Un Bite); Marta Gonzalez (Ocean Deli); Arturo Carrion and Karlox Ayala (Peko Peko); High Kitchen; Lemon Submarine; Pisco Labis. You know I love you all! You were the core of this operation from day one. You were our ears and eyes when it came to the needs of the people. You will be my family forever.

To the kitchens we gratefully borrowed: José Enrique in Santurce, Centro Envejecientes in Vieques, Head Start in Utuado, Vivo Beach Club in Carolina, José Miguel Agrelot Coliseum (El Choliseo).

To our biggest suppliers and supporters: Mario Pagan, one of the first people I called, who introduced me to his old friend Jorge Unanue, who was so generous donating Goya food and his helicopter pilot skills to take us to the most unreachable parts of Puerto Rico; Ramón Leal of ASORE, my Puerto Rican brother and anchor, who gave me the credibility I needed and who worked harder than anybody in the early days and weeks; Ramón Gonzalez Cordero of Empire Gas, who made sure we had gas in every situation, and worked tirelessly to get gas to anybody who needed it across the island; Ramón Santiago and Eduardo Santiago from José Santiago, my fellow Asturians and our favorite food providers, as well as the perfect example of how the private sector works to perfection; Alberto de la Cruz of Coca-Cola, who provided us with water, know-how and crucial connections, and who had the clearest solutions to the gas and water problems on the island; Viviana Mercardo of Walmart, who helped us skip the lines at Sam's Club, and gave us help with food and money early on when nobody else believed in us; Lulu Puras and Guillin Arzuaga of Mano y Mano, who did so much to deliver food to the *ejidas* and supported chef Piñeiro from the earliest days; Bernardo Medina, our media and communications expert, who helped me navigate the media waters of Puerto Rico in a smooth way and helped us with the branding that is more important than you think because it gives you a sense that we are one team; José Luis Labeaga of Mi Pan bakery, who was so important in our sandwich operation, alongside the island's other bakeries; Roberto Cacho, who helped us understand the power of water filters and gave us huge support early on. And to Andrés López for all your wise legal advice.

Our thanks to Carlos Vazquez and the whole team at The Place in Condado, for feeding our team almost every night: thank you for making us feel at home when we were away from home and for giving us a place to belong.

And our special thanks to the whole team at the AC Hotel, which was our home for many months. They were always patient, understanding and supportive, even when we changed our plans late at night in a building that was struggling with power and over-booking.

To my Dorado Beach team, including Friedel and Federico Stubbe of the Prisa Group, who are my friends and family, and my secret weapon to get anything done; Kenny Blatt, for always caring for me and helping us use the kitchen at Dorado Beach; Arne Sorenson, CEO of Marriott International, for all your support.

To my friend and partner Fedele Bauccio, from Bon Appétit, who sent chefs down to support our efforts, without asking why. I now understand why his company is one of the best food operations in the world. And to Gary Green and all my friends at the Compass Group, whose many companies helped in so many ways. Also to Restaurant Associates for all their help in Houston.

To Emilio and Gloria Estefan, who will always be my brother and my sister, and who helped Puerto Rico in so many generous ways.

To Daya Fernandez, for calling me from Paris with guidance about the places that needed food in Puerto Rico and for the connection to Lulu Puras; David Naranjo at Rock Orange, for offering help with planes and so much more; Lymari Nadal, for showing up at the beginning and bringing so many meals to Ponce; Luis Fortuño, the former governor of Puerto Rico, for all your support and advice; Senator Mark Warner, for caring and calling and asking how we needed help.

To Jimmy Kemp of the Jack Kemp Foundation: thank you for your great advice and connections in Congress. To the Fonalledas family, for getting us products, assistance and moral support early on. To Juan Carlos Iturregui, for being such a great voice of wisdom.

Our thanks to FreshPoint for donating vegetables and especially a refrigerator truck for weeks on end. We couldn't have ramped up in the early days and weeks without you. Thanks also to Mario Somoza from Pan Pepín, for all the food you donated; and to the IRSI restaurant group for donating money and food to support us.

To the many mayors and senators in Puerto Rico: thank you for your devotion to helping people and your support for us. You know who you are, but you are too many to mention and my gratitude is no less for all that.

My thanks and love to Lin-Manuel Miranda, because you are you, and because a humble rap tweet gave us the energy we needed to keep going at a tough time.

I cannot thank enough Laurene Powell Jobs: because nobody does more without expecting anything in return. You are my sister, and you gave me hope in a moment of darkness in more ways than you will ever imagine. Also my thanks to Stacey Rubin, managing director of Emerson Collective, who believed in us all along, was always just a phone call away, and who opened the doors of impossibility.

To Ted Leonsis and Zack Leonsis and all the Monumental family: nobody cares more about community in Washington than you. And when Hurricane Maria happened, nobody came to our aid quicker than you did.

To Fred and Karen Schaufeld, for caring for me at any distance, for making sure my well-being was taken care of, and for making my life easier when returning home. To Herb Allen, for all your support in easing my return home, and to Jeff Bezos for answering my calls for help.

In Puerto Rico's government, my thanks to: Beatriz Rosselló, the first lady, whose Stop and Go initiative we were proud to partner with; Leila Santiago in the first lady's office, for always answering the phone when we needed help; Julia Keleher, for being an education secretary who tried to use the schools to feed the people; Secretary of State Luis Rivera-Marín, for supporting our feeding operation and being responsive whenever we called; the mayor of Ponce, Maria "Mayita" Melendez Altieri, who became a good friend, and who gave us early support to help her community.

In the media, my thanks to so many journalists who were just doing their job, which was so much more than that. Especially to the forgotten photojournalists who gave us clues about where we needed to deliver food across the island. To David Begnaud of CBS News, Bill Weir of CNN, and Anderson Cooper of CNN and 60 *Minutes*: thank you for keeping us all informed, and for helping us feed the people because of your reporting. To Kim Severson of the *New York Times*, Jorge Ramos of Univision, Robin Roberts of ABC's *Good Morning America*, photographer Eric Rojas, Rubén Sánchez of Univision, and political analyst Jay Fonseca: thank you for caring and taking the time to do such a great job every day.

To our partners at Mercy Corps, especially Javier Alvarez, Jeronimo Candela and Neal Keny-Guyer: thank you for being agents of change. To my friends at Oxfam, for listening to us and giving us some direction. To Luis and Frankie Miranda at the Hispanic Federation: thank you for all the amazing work you do and for giving us credibility with your support. To Unidos Por Puerto Rico: thank you to the entire board for the amazing work you did, and for visiting and supporting our operations.

To Juan Del Rio Martin of the Auxilio Mutuo hospital, for doing heroic transplant work in the worst conditions after the hurricane; to Jeffrey Akman of George Washington University, for delivering life-saving medicines; to the National Guard in Puerto Rico, who helped us complete our missions without any planning or warning; to the many volunteers from the United States Coast Guard; to the map-makers from the U.S. Army Corps of Engineers, whose skills are under-appreciated but still incredibly valuable; and to the men and women of FEMA, who were away from their families for weeks on end, trying to achieve good for Americans in need.

My special thanks to the many agents of Homeland Security Investigations, especially Assistant Special Agent In Charge (ASAC) Jorge Uribarri from El Paso, for helping me find a close family member, and for delivering food and water to some of the most remote and difficult areas from the earliest days of our operation; to ASAC Bernardo Pillot from Atlanta; and to Group Supervisors Andres Maldonado, Michal Lopez, Walter Rivera, Mike Ortiz, Jerry Conrad, James Clark, Rex Setzer, and Ritchie Flores.

To my World Central Kitchen family for believing in our mission: Brian MacNair, Kevin Holst, and Jeanette Morelan. And to our World Central Kitchen board for being so supportive: Fredes Montes, Javier Garcia, Jean Marc DeMatteis, Kevin Doyle, Lizette Corro, Rob Wilder, Victor Albisu. Especially to Robert Egger, who showed me how food can change the world, and who was my early mentor on all things to do with improving lives one meal at a time.

Also to Mike Curtin, who has been leading the way in D.C. with the most forward-thinking NGO I know. To Patty Stonesifer, president and CEO of Martha's Table, who has been so visionary in the nonprofit world. And to Bill Shore, because it was at Share Our Strength that I began learning the power of sharing knowledge with many women in D.C. through the Frontline classes.

To Guy Fieri, for activating a food truck in the Napa fires: we are family now. To David Chang, for supporting us in Ventura and helping us raise money. To Eric and Sandra Ripert, for caring for Puerto Rico and looking after me from afar. And to Stan Hayes from Operation Barbecue Relief, for doing amazing work in Houston and elsewhere.

To my doctor Kevin Kelleher for coming to my home at a moment's notice to fix me.

To my beloved agent Kimberly Witherspoon, who keeps pushing books right and left on my behalf and is always a good voice representing me like nobody else. Also to my friend Kris Dahl, who represents my friend Richard Wolffe and always has great words of wisdom.

To my buddy Anthony Bourdain, who with the power of a tweet showed us early support and gave us a strong voice on many issues. To Dan Halpern at Ecco who believed without any hesitation that I could not only give him on time a vegetable book but also this amazing story of Puerto Rico. And to Matt Goulding for being patient and putting up with me as I let him down in the process of writing the vegetable book, and for writing my op-ed on Catalonia in the middle of everything.

To my friend of the last twenty years Richard Wolffe, who has written all my bullshit all my life, and who came to be next to me without asking why. And to his family Paula, Ilana, Ben and Max, for following closely everything that was happening, and for showing up at Thanksgiving to help and support us in Puerto Rico.

Above all, to my family: Nothing would be possible without a supportive family. To my wife Patricia Fernandez de la Cruz, who is caring, loving and the best friend a nutjob like me can have: I look forward to watching more sunsets with you until the end of my life. And to our daughters Carlota, Ines, and Lucia, who wished they were in Puerto Rico throughout, and showed up at Thanksgiving to help prepare and serve more than 40,000 meals. It was the best Thanksgiving we ever had.

NOTES

Prologue

1. National Constitution Center, "How a Hurricane Brought Alexander Hamilton to America," 8/31/17, https://constitutioncenter.org/blog/how-a-hurricane-brought-an-important-founding-father-to-america.
2. Letter from Alexander Hamilton to The Royal Danish American Gazette, 9/6/1772, accessed at the National Archives website: https://founders.archives.gov/documents/Hamilton/01-01-02-0042.
3. *Guardian*, "Trump Attempts to Use Spanish Accent to Pronounce Puerto Rico—Video," 8/6/17, https://www.theguardian.com/us-news/video/2017/oct/07/trump-attempts-to-use-spanish-accent-to-pronounce-puerto-rico-video.

Chapter 1: Landfall

1. Pritha Paul, "Hurricane Maria, Now Category 5, Blows Away Roof of Dominica PM's House," *IBT*, 9/19/17, http://www.ibtimes.com/hurricane-maria-now-category-5-blows-away-roof-dominica-pms-house-2591299.
2. Mattathias Schwartz, "100 Days of Darkness," *New York*, 12/25/17.
3. Remarks By President Trump and President Poroshenko of Ukraine Before Bilateral Meeting, Lotte New York Palace Hotel, transcribed by The White House, 9/21/17, https://www.whitehouse.gov/briefings-statements/remarks-president-trump-president-poroshenko-ukraine-bilateral-meeting/.
4. Abby Phillip, Ed O'Keefe, Nick Miroff and Damian Paletta, "Lost Weekend: How Trump's Time At His Golf Club Hurt the Response to Maria," *Washington Post*, 9/29/17, https://www.washingtonpost.com/politics/lost-weekend-how-trumps-time-at-his-golf-club-hurt-the-response-to-maria/2017/09/29/ce92ed0a-a522-11e7-8c37-e1d99ad6aa22_story.html?utm_term=.322d89eccc63.
5. Christopher Gillette, "Aid Begins to Flow to Hurricane-Hit Puerto Rico,"

Associated Press, 9/24/17, https://apnews.com/06f5077aff384e508e2f23 24dae4eb2e.

6. Terri Moon Cronk, "DoD Continues Round-the-Clock Support Following Hurricanes in Caribbean," *DoD News*, 9/25/17, https://www .defense.gov/News/Article/Article/1323530/dod-continues-round-the -clock-support-following-hurricanes-in-caribbean/.

7. Deirdre Walsh and Kevin Liptak, "Federal Response to Hurricane Maria Slowly Takes Shape," *CNN*, 9/25/17, http://www.cnn.com /2017/09/25/politics/puerto-rico-hurricane-maria-aid-donald-trump /index.html.

8. Donald Trump (@realDonaldTrump), *Twitter*, 9/25/17. https://twitter .com/realdonaldtrump/status/912479500511965184?lang=en.

9. Abby Phillip, Ed O'Keefe, Nick Miroff and Damian Paletta.

10. Barbara Starr, Zachary Cohen and Ryan Browne, "US Military Sends Ships, Aircraft to Puerto Rico," *CNN*, 9/26/17 http://www.cnn.com /2017/09/26/politics/us-military-response-puerto-rico-hurricane -maria/index.html.

11. Marco Rubio and Bill Nelson, Letter to President Donald Trump, 9/26/17, https://www.rubio.senate.gov/public/_cache/files/9e20326b -8beb-4363-acbb-a2bf4114bb63/478FE40F28C96847538E284DB334066C .17.09.26-smr-letter-to-potus-re-pr-w.-signatures.pdf.

12. Abby Phillip, Ed O'Keefe, Nick Miroff and Damian Paletta.

13. "DoD, Partner Agencies Support Puerto Rico, Virgin Islands Hurricane Relief Efforts," *DoD News*, 9/26/17, https://www.defense.gov/News /Article/Article/1325245/dod-partner-agencies-support-puerto-rico -virgin-islands-hurricane-relief-efforts/.

14. Mattathias Schwartz.

15. Olga Khazan, "The Crisis at Puerto Rico's Hospitals," *The Atlantic*, 9/26/17, https://www.theatlantic.com/health/archive/2017/09/the -crisis-at-puerto-ricos-hospitals/541131/.

16. Omaya Sosa Pascual, "Hurricane Maria's Death Toll in Puerto Rico Is Higher Than Official Count, Experts Say," *Miami Herald*, 9/28/17, http:// www.miamiherald.com/news/weather/hurricane/article175955031 .html.

Chapter 2: Feed the World

1. Jerry Adler, "Why Fire Makes Us Human," *Smithsonian*, 6/13, https:// www.smithsonianmag.com/science-nature/why-fire-makes-us -human-72989884/.

2. Jonathan M. Katz, *The Big Truck That Went By: How The World Came To Save Haiti And Left Behind A Disaster* (St. Martin's Griffin, 2013: pp. 35-52)

3. Ibid, pp. 67-86.

4. Ibid, p. 239.

5. Ibid, pp. 217-44.

6. Ibid, pp. 67-86.

7. Ibid, pp. 204-5.

8. Ibid, p. 106.

9. Ibid, pp. 206-7.
10. Ibid, p. 146.
11. Michele Landis Dauber, *The Sympathetic State: Disaster Relief and the Origins of the American Welfare State* (University of Chicago Press, 2013: pp. 17-23).
12. Ibid, pp. 79-125.
13. FEMA video, "Southern Baptist Disaster Relief," 6/15/13, https://www.fema.gov/media-library/assets/videos/82819#embed-code.
14. Sarah Eekhoff Zylstra, "How Southern Baptists Trained More Disaster Relief Volunteers Than The Red Cross," *The Gospel Coalition*, 11/17/17, https://www.thegospelcoalition.org/article/how-southern-baptists-trained-more-disaster-relief-volunteers-than-the-red-cross/.
15. Southern Baptist Disaster Relief website, https://www.namb.net/send-relief/disaster-relief.

Chapter 3: Discovery
1. Jorge Duany, *Puerto Rico: What Everyone Needs To Know* (Oxford University Press, 2017: pp. 9-12).
2. Ibid, pp. 13-19.
3. Ibid, pp. 24-29.
4. Ibid, pp. 30-33.
5. Ibid, pp. 40-44.
6. Ibid, pp. 44-49.
7. Ibid, pp. 92-98.
8. Ibid, pp. 61-63.
9. Ibid, pp. 81-82.
10. Frances Robles, "23% Of Puerto Ricans Vote in Referendum, 97% of Them for Statehood," *New York Times*, 6/11/17, https://www.nytimes.com/2017/06/11/us/puerto-ricans-vote-on-the-question-of-statehood.html.
11. United States Census Bureau, QuickFacts: Puerto Rico https://www.census.gov/quickfacts/PR.
12. Carmen Sesin, "Over 200,000 Puerto Ricans Have Arrived in Florida Since Hurricane Maria," *NBC News*, 11/30/17, https://www.nbcnews.com/news/latino/over-200-000-puerto-ricans-have-arrived-florida-hurricane-maria-n825111.
13. John D. Sutter and Sergio Hernandez, "Exodus From Puerto Rico: A Visual Guide," *CNN*, 2/21/18, https://www.cnn.com/2018/02/21/us/puerto-rico-migration-data-invs/index.html.
14. Jorge Duany, p. 3.
15. Ibid, p.163.
16. Elizabeth Wolkomir, "How Is Food Assistance Different in Puerto Rico Than in the Rest of the United States?" Center on Budget and Policy Priorities, 11/27/17, https://www.cbpp.org/research/food-assistance/how-is-food-assistance-different-in-puerto-rico-than-in-the-rest-of-the.
17. Cruz Miguel Ortíz Cuadra, *Eating Puerto Rico: A History Of Food, Culture, And Identity* (University of North Carolina Press, 2013: pp. 17-21).

18. Ibid, p 31.

19. Ibid, p. 149.

20. Ibid, p. 233.

Chapter 4: Big Water

1. Press Briefing by Press Secretary Sarah Sanders, transcribed by The White House, 9/28/17, https://www.whitehouse.gov/briefings -statements/press-briefing-press-secretary-sarah-sanders-092817/.

2. Omaya Sosa Pascual, "Hurricane Maria's Death Toll in Puerto Rico Is Higher Than Official Count, Experts Say," *Miami Herald*, 9/28/17, http:// www.miamiherald.com/news/weather/hurricane/article175955031 .html.

3. Danny Vinik, "How Trump Favored Texas Over Puerto Rico," *Politico*, 3/27/18, https://www.politico.com/story/2018/03/27/donald-trump -fema-hurricane-maria-response-480557.

4. Report on the Competitiveness of Puerto Rico's Economy, Federal Reserve Bank of New York, 6/29/12, https://www.newyorkfed.org /medialibrary/media/regional/PuertoRico/report.pdf.

5. Puerto Rico: Characteristics of the Island's Maritime Trade and Potential Effects of Modifying The Jones Act, United States. Government Accountability Office, March 2013, https://www.gao.gov /assets/660/653046.pdf.

6. Ron Nixon and Matt Stevens, "Harvey, Irma, Maria: Trump Administration's Response Compared,"*New York Times*, 9/27/17, https://www.nytimes.com/2017/09/27/us/politics/trump-puerto-rico -aid.html.

7. Press Gaggle by President Trump, transcribed by The White House, 9/27/17, https://www.whitehouse.gov/briefings-statements/press -gaggle-president-trump/.

8. Natalie Andrews and Paul Page, "Trump Weights Waiving Law Barring Foreign Ships From Delivering Aid to Puerto Rico," *Wall Street Journal*, 9/27/17, https://www.wsj.com/articles/lawmakers-seek-waiver-of -law-barring-foreign-ships-from-delivering-aid-to-puerto-rico -1506529999.

9. Niraj Chokshi, "Trump Waives Jones Act for Puerto Rico, Easing Hurricane Aid Shipments," *New York Times*, 9/28/17, https://www .nytimes.com/2017/09/28/us/jones-act-waived.html?_r=0.

10. American Red Cross/FEMA Memorandum of Agreement Implementation Meeting Summary Notes, 1/21/11, https://nmcs .communityos.org/cms/files///NMCS%20Meeting%201-21-11%20 Overview%20Notes.pdf.

11. National Mass Care Strategy: A Roadmap for the National Mass Care Service Delivery System, September 2012, http://www .nationalmasscarestrategy.org/wp-content/uploads/2014/07/national -mass-care-strategy-september-2012-_comp.pdf.

12. Remarks by President Trump to the National Association of Manufacturers, transcribed by The White House, 9/29/17, https://

www.whitehouse.gov/briefings-statements/remarks-president
-trump-national-association-manufacturers/.

13. Amanda Holpuch, "San Juan Mayor's Harrowing Plea: 'Mr Trump, I Am
Begging. We Are Dying Here.'" *Guardian*, 9/29/17, https://www
.theguardian.com/world/2017/sep/29/san-juan-mayor-plea-donald
-trump-puerto-rico.

14. Abby Phillip, Ed O'Keefe, Nick Miroff and Damian Paletta, "Lost
Weekend: How Trump's Time at His Golf Club Hurt the Response to
Maria," *Washington Post*, 9/29/17, https://www.washingtonpost.com
/politics/lost-weekend-how-trumps-time-at-his-golf-club-hurt
-the-response-to-maria/2017/09/29/ce92ed0a-a522-11e7-8c37
-e1d99ad6aa22_story.html?utm_term=.10d7e31cc986.

15. Juana Summers, "Trump Attacks San Juan Mayor Over Hurricane
Response," CNN, 9/30/17, https://www.cnn.com/2017/09/30/politics
/trump-tweets-puerto-rico-mayor/index.html.

16. Jim Garamone, "DoD Boosts Personnel Aiding Hurricane Relief Efforts
In Puerto Rico," *DoD News*, 10/1/17, https://www.defense.gov/News
/Article/Article/1330602/dod-boosts-personnel-aiding-hurricane
-relief-efforts-in-puerto-rico/.

17. "DoD Accelerates Hurricane Relief, Response Efforts in Puerto Rico,"
DoD News, 9/30/17, https://www.defense.gov/News/Article/Article
/1330501/dod-accelerates-hurricane-relief-response-efforts-in
-puerto-rico/.

18. "FEMA Administrator: Puerto Rico Has 'a Long Way to Go' on
Hurricane Recovery," *ABC News*, 10/1/17, http://abcnews.go.com
/ThisWeek/video/fema-administrator-puerto-rico-recovery-relief
-efforts-50208624.

Chapter 5: In the Arena

1. Omaya Sosa Pascual, "Pesquera Reconoce Que Hay Más Muertos
Por María," *Centro de Periodismo Investigativo*, 10/2/17, http://
periodismoinvestigativo.com/2017/10/pesquera-reconoce-que-hay
-mas-muertos-por-maria/.

2. Jim Garamone, "Unified Coordination Group Addressing Puerto Rico
Recovery, Governor Says," *DoD News*, 10/2/17, https://www.defense
.gov/News/Article/Article/1331056/unified-coordination-group
-addressing-puerto-rico-recovery-governor-says/.

3. Mark Landler, "Trump Lobs Praise, and Paper Towels, to Puerto Rico
Storm Victims," *New York Times*, 10/3/17, https://www.nytimes.com
/2017/10/03/us/puerto-rico-trump-hurricane.html.

4. Remarks by President Trump in Briefing on Hurricane Maria Relief
Efforts, transcribed by The White House, 10/3/17, https://www
.whitehouse.gov/briefings-statements/remarks-president-trump
-briefing-hurricane-maria-relief-efforts/.

5. Caroline Kenny, "Trump Tosses Paper Towels Into Puerto Rico Crowd,"
CNN, 10/3/17, https://www.cnn.com/2017/10/03/politics/donald
-trump-paper-towels-puerto-rico/index.html.

6. "US Military Beefs Up Its Efforts in Puerto Rico as Need for Vital Supplies Grows," *CBS News*, 9/28/17, https://www.cbsnews.com/news/military-efforts-puerto-rico-hurricane-maria/.

7. "USS *Kearsarge* (LH3) Providing Critical Hurricane Relief," *Navy Supply Corps Newsletter*, 2/6/18, http://scnewsltr.dodlive.mil/2018/02/06/uss-kearsarge-lhd-3-providing-critical-hurricane-relief/.

8. "Update: US Northern Command Continues Humanitarian Aid to Puerto Rico and the US Virgin Islands," Press Release by US Northern Command, 9/25/17, http://www.northcom.mil/Newsroom/Press-Releases/Article/1323710/update-us-northern-command-continues-humanitarian-aid-to-puerto-rico-and-the-us/.

9. Mary Williams Walsh and Alan Rappeport, "White House Dials Back Trump's Vow to Clear Puerto Rico's Debt," *New York Times*, 10/4/17, https://www.nytimes.com/2017/10/04/business/dealbook/trump-puerto-rico-debt.html.

10. Press Gaggle by President Trump, Press Secretary Sarah Sanders, Congresswoman Jenniffer Gonzalez-Colon, and Small Business Administrator Linda McMahon, transcribed by The White House, 10/3/17, https://www.whitehouse.gov/briefings-statements/press-gaggle-president-trump-press-secretary-sarah-sanders-congresswoman-jenniffer-gonzalez-colon-small-business-administrator-linda-mcmahon/.

11. David Begnaud, "Woman Behind Botched FEMA Contract to Deliver Meals in Puerto Rico Speaks Out," *CBS News*, 2/8/18, https://www.cbsnews.com/news/woman-behind-botched-fema-contract-to-deliver-meals-in-puerto-rico-speaks-out/.

12. Patricia Mazzei and Agustin Armendariz, "FEMA Contract Called for 30 Million Meals for Puerto Ricans. 50,000 Were Delivered," *New York Times*, 2/6/18, https://www.nytimes.com/2018/02/06/us/fema-contract-puerto-rico.html.

13. Tami Abdollah, "AP Exclusive: Big Contracts, No Storm Tarps for Puerto Rico," *Associated Press*, 11/28/17, https://www.apnews.com/cbeff1a939324610b7a02b88f30eafbb.

14. Letter to Trey Gowdy, Chairman of the House Committee on Oversight and Government Reform, from Elijah Cummings, Ranking Member, and Stacey Plaskett, Member of Congress, 2/6/18, https://democrats-oversight.house.gov/sites/democrats.oversight.house.gov/files/2018-02-06.EEC%20%26%20Plaskett%20to%20Gowdy%20re.FEMA-Tribute%20Contracting.pdf.

15. Ibid.

16. Ken Klippenstein, "$300M Puerto Rico Recovery Contract Awarded to Tiny Utility Company Linked to Major Trump Donor," *Daily Beast*, 10/24/17, https://www.thedailybeast.com/dollar300m-puerto-rico-recovery-contract-awarded-to-tiny-utility-company-linked-to-major-trump-donor.

17. Frances Robles, "The Lineman Got $63 an Hour. The Utility Was Billed $319 an Hour," *New York Times*, 11/12/17, https://www.nytimes.com

/2017/11/12/us/whitefish-energy-holdings-prepa-hurricane-recovery
-corruption-hurricane-recovery-in-puerto-rico.html.

18. Frances Robles and Deborah Acosta, "Puerto Rico Cancels Whitefish Energy Contract to Rebuild Power Lines," *New York Times*, 10/29/17, https://www.nytimes.com/2017/10/29/us/whitefish-cancel-puerto -rico.html.

19. Frances Robles, "Puerto Rico's Health Care Is in Dire Condition, Three Weeks After Maria," *New York Times*, 10/10/17, https://www.nytimes .com/2017/10/10/us/puerto-rico-power-hospitals.html.

20. "EPA Hurricane Maria Update for Wednesday, October 11th," Press Release by EPA, 10/11/17, https://www.epa.gov/newsreleases/epa -hurricane-maria-update-wednesday-october-11th.

21. John Sutter, "EPA: Water at Puerto Rico Superfund Site Is Fit for Consumption," *CNN*, 10/31/17, https://www.cnn.com/2017/10/31 /health/puerto-rico-water-epa-superfund-test-results/index.html.

22. Mary Williams Walsh and Alan Rappeport, "White House Dials Back Trump's Vow to Clear Puerto Rico's Debt," *New York Times*, 10/4/17, https://www.nytimes.com/2017/10/04/business/dealbook/trump -puerto-rico-debt.html.

23. Luis Ferré-Sadurní, "Higher Puerto Rico Death Toll Reflects Survey Across Island," *New York Times*, 10/4/17, https://www.nytimes.com /2017/10/04/us/puerto-rico-death-toll-maria.html.

Chapter 6: Ready to Eat

1. Lisa Burgess, "MRES: It Could Be Worse (And It Was)," *Stars and Stripes,* 3/16/08, https://www.stripes.com/lifestyle/mres-it-could-be-worse -and-it-was-1.77097.

2. Peggy Mihelich, "Grub, Chow, Mystery Meat – Combat Food 2.0," *CNN*, 9/13/07, Foodhttp://www.cnn.com/2007/TECH/09/13/combat.food /index.html.

3. "Meal, Ready-To-Eat," Webpage by Defense Logistics Agency, http:// www.dla.mil/TroopSupport/Subsistence/Operationalrations/mre.aspx.

4. Nathaniel Weixel, "Trump Officials Allow Puerto Ricans to Use Food Stamps for Hot Food," *The Hill,* 10/3/17, http://thehill.com/blogs/blog -briefing-room/353685-trump-admin-denied-puerto-rico-request-to -let-hurricane-victims-use.

5. "Update: US Northern Command Continues Puerto Rican Relief Efforts," Press Release by US Northern Command, 10/9/17, http:// www.northcom.mil/Newsroom/Press-Releases/Article/1337668 /update-us-northern-command-continues-puerto-rican-relief-efforts/.

Chapter 7: Seeing Red

1. "FEMA Expands Leadership Team in Puerto Rico," Press Release by FEMA, 10/11/17, https://www.fema.gov/news-release/2017/10/11 /fema-expands-leadership-team-puerto-rico.

2. "Can I Quit Now? FEMA Chief Wrote As Katrina Raged," *CNN*, 11/4/05, http://www.cnn.com/2005/US/11/03/brown.fema.emails/.

3. "Speaker Ryan Visits Puerto Rico, Pledges Continued Support," Press Release by Speaker Ryan Press Office, 10/13/17, https://www.speaker .gov/press-release/speaker-ryan-visits-puerto-rico-pledges-continued -support.

4. Alexia Fernández Campbell, "Puerto Rican Officials Claim the Water Crisis Is Under Control. Reports on the Ground Tell a Very Different Story," Vox, 10/25/17, https://www.vox.com/policy-and-politics /2017/10/25/16504870/puerto-rico-running-water.

5. Hurricane Maria Three-Month Update, Webpage by the American Red Cross, January 2018, http://embed.widencdn.net/pdf/plus /americanredcross/vfimlniity/hurr-maria-three-month-update .pdf?u=0aormr&proxy=true.

Chapter 8: Transitions

1. Tim Carman, "After Maria, José Andrés and His Team Have Prepared More Hot Meals in Puerto Rico Than the Red Cross," Washington Post, 10/18/17, https://www.washingtonpost.com/news/food/wp/2017/10/18 /post-maria-jose-andres-and-his-team-have-served-more-meals-in -puerto-rico-than-the-red-cross/?utm_term=.7b83eae80581.

2. Jake Gibson, "FBI in Puerto Rico Investigating If Corrupt Local Officials Are 'Withholding' or 'Mishandling' Crucial Supplies," Fox News, 10/12/17, http://www.foxnews.com/us/2017/10/12/fbi-in-puerto-rico -investigating-mishandling-fema-supplies.html.

3. "An Update to US Northern Command's Support to Puerto Rico," Press Release by US Northern Command, 10/19/17, http://www.northcom .mil/Newsroom/Press-Releases/Article/1349856/an-update-to-us -northern-commands-support-to-puerto-rico/.

4. National Constitution Center, "How a Hurricane Brought Alexander Hamilton to America," 8/31/17, https://constitutioncenter.org/blog /how-a-hurricane-brought-an-important-founding-father-to-america.

5. Adrian Carrasquillo, "Chef José Andrés and the Trump Administration Are Fighting Over Puerto Rico," BuzzFeed, 11/6/17, https://www .buzzfeed.com/adriancarrasquillo/chef-José-andres-and-the-trump -administration-are-fighting?utm_term=.cyqoaoNeW#.gp4NVNmJo.

6. Mahita Gajanan, "The American Government Has Failed." Celebrity Chef José Andrés Slams FEMA's Puerto Rico Response," TIME, 10/16/17, http://time.com/4981655/José-andres-fema-trump-puerto-rico/.

Epilogue

1. Tim Sullivan, "Embracing Complexity," Harvard Business Review, September 2011, https://hbr.org/2011/09/embracing-complexity.

2. "Keeping Faith With Our Fellow Americans" Alice Thomas, Refugees International, December 2017.

3. Jonathan M. Katz, The Big Truck That Went By: How The World Came To Save Haiti And Left Behind A Disaster (St. Martin's Griffin, 2013: p. 105).

4. "Using Checklists to Prevent Failure," Harvard Business Review, January 2010, https://hbr.org/2010/01/using-checklists-to-prevent-fa.html.

5. Richard Parker, "The Military Was Ready in Texas and Florida. What Went Wrong in Puerto Rico?" *Politico*, 10/2/17, https://www.politico.com/magazine/story/2017/10/02/us-military-puerto-rico-215668.

6. Tami Abdollah, "Emails Show FEMA Silent As Puerto Rico Sought Generator Fuel," *Washington Post*, 3/21/18, https://www.washingtonpost.com/business/emails-show-fema-silent-as-puerto-rico-sought-generator-fuel/2018/03/21/29286352-2cbf-11e8-8dc9-3b51e028b845_story.html?utm_term=.cf0aeb6269a9.

7. Testimony of Brock Long, FEMA administrator, under questioning by the House Homeland Security Committee, "Preparedness, Response and Rebuilding: Lessons from the 2017 Disasters," 3/15/18, https://www.youtube.com/watch?v=GRA51-EZX58.

8. Oxfam's country director in Haiti admitted using prostitutes at his Oxfam villa, and was allowed to resign without disciplinary action. See Oxfam's official report here: "Oxfam Releases Report Into Allegations of Sexual Misconduct in Haiti," Press Release by Oxfam, 2/19/18, https://www.oxfam.org/en/pressroom/pressreleases/2018-02-19/oxfam-releases-report-allegations-sexual-misconduct-haiti.

9. Justin Elliott, Jesse Eisinger and Laura Sullivan, "The Red Cross' Secret Disaster," *ProPublica*, 10/29/14, https://www.propublica.org/article/the-red-cross-secret-disaster.

10. Letter from Richard Rieckenberg to Trevor Riggen, 11/18/12, https://www.documentcloud.org/documents/1346529-letter-to-trevor-riggen-nov-18-2012.html#document/p1/a184560.

11. Nishant Kishore *et al.*, "Mortality in Puerto Rice after Hurricane Maria," *The New England Journal of Medicine*, 5/29/18.

12. Taylor Rock, "José Andrés Is Feeding Californians Displaced by Violent Wildfires," *Los Angeles Times*, 12/7/17, http://www.latimes.com/food/sns-dailymeal-1860436-eat-jose-andres-feeds-california-fire-120717-20171207-story.html.